'Wonderfully enlightening and engagingly written. "Well Fed: How the way we eat is destroying our health and planet (and how to fix it)" navigates the complex world of food systems with clarity and thoughtful insight. James skillfully demystifies nutritional science and sustainability, guiding readers through the imperfections of our food system and the importance of informed food choices.'

Dr Idz

'James combines decades of experience as a nutritionist, a wide range of expert sources with a deep understanding of how we assimilate and digest information today to bring all important mindfulness back into what we eat.'

Mallika Basu

'If you are looking at ways to combat the exploitation of the planet this book is for you. Outlining the environmental, nutritional and ethical considerations it is possible to contribute towards the change we desperately need to see. James breaks down the science into attainable actions, simple changes to our diets that are necessary and achievable. Thank you for bringing the ethics of the food we are so deeply disconnected from to the forefront of our minds.'

Rhiannon Lambert

'In his insightful book, James has skilfully shown us the links between human health, human ethics and planetary health like never before. He highlights the importance of combining traditions (home cooked meals and eating with others) with modern nutrition science and food technology to provide a hopeful vision for a healthy future for humans and planet.'

Dr Gemma Newman

'James Collier's book *Well Fed* brings a fresh perspective to the complex relationship between nutrition, our health, and the planet. His approach empowers readers to make informed choices, balancing personal well-being with environmental sustainability. As a fellow advocate for reducing (less nutritious) ultra-processed foods, I fully support his vision of a food system where eating well also means caring for our planet.'

Nichola Ludlam-Raine

'*Well Fed* is a fresh take on how our everyday food choices shape our health and the planet, that's both thought-provoking and practical.'

Dr Emily Leeming

'An insightful and upbeat book on the history, ethics and sustainability of food. James shows us how we ended up with the challenges of our modern food landscape, and gives us the tools to navigate it.'

Barbara Bray MBE

'James Collier's *Well Fed* is a groundbreaking exploration of our food choices and their far-reaching impacts on health, ethics, and the environment. With clarity and compassion, he guides us through the complexities of the modern food system, encouraging thoughtful reflection and empowering readers to make informed decisions. This book is a must-read for anyone looking to enhance their well-being while contributing to a more sustainable planet.'

Kristen Holmes

JAMES COLLIER

How modern diets are failing us (and what we can do about it)

Well Fed

Thorsons

Thorsons
An imprint of HarperCollins*Publishers*
1 London Bridge Street
London SE1 9GF

www.harpercollins.co.uk

HarperCollins*Publishers*
Macken House, 39/40 Mayor Street Upper
Dublin 1, D01 C9W8, Ireland

First published by Thorsons 2025

10 9 8 7 6 5 4 3 2 1

A catalogue record of this book is
available from the British Library

HB ISBN 978-0-00-873236-3
PB ISBN 978-0-00-873766-5

Printed and bound in the UK using 100%
renewable electricity at CPI Group (UK) Ltd

For my mother, Marilynne Collier

Contents

Part I

Contemplating the Food System

Chapter 1

An Imperfect Food System

'The global food system is broken.'

I often hear nutrition and climate professionals make this statement. They have a point. After all, with worsening diets, environmental pressures and questionable agricultural practices, it's as crucial as it is urgent that our food system is improved. It may surprise you, however, to learn that I don't wholly agree with the claim. Our food system is not broken, but it is far from perfect. Indeed, in many ways, over the past several decades, it's been getting better. For example, more and more people have greater access to sufficient calories and protein: the number of people facing hunger decreased by 150 million globally between 2005 and 2019.[1] Yet, in other respects, it's travelling in the wrong direction: our environment is being destroyed, animals are increasingly exploited and metabolic diseases of excess prevail.

The number of people afflicted with the numerous health conditions relating to diet and lifestyle continues to rise. With people continuing to fall ill as a result of what they eat, we're increasingly reliant on the advances of medicine to overcome the wrath of poor nutrition, when, from one perspective, many of the food and life-style-related illnesses shouldn't be an issue in the first place. The modern food system is a major contributor to greenhouse gas emissions, worsening the critical nature of the climate crisis. Despite the fact that there's plenty of food to comfortably feed every human alive

– so much so that more than a billion tonnes of food are wasted every year[2] – undernutrition continues to remain a problem for millions of people. The way we grow our crops makes them less nutritious than nature originally intended, yet people who would otherwise struggle to provide for their families rely on mass growing for sustenance. There are justifiable concerns about the suffering inflicted on animals through intensive farming practices, some of which also contribute to poorer quality of food and environmental destruction. If we continue to fish in the way we do, it will not be long before we disrupt the whole marine ecosystem, leaving insufficient fish left available in the seas. In addition, poor food choice is linked to many of the problems people are experiencing with their mental health.

These are just some of the key concerns relating to our imperfect food system. All of them can be resolved. Each issue demonstrates the imperfections of our ever-expanding food system. While some of the problems have been driven by economic incentives, greed and shoddy political decisions, many are the result of good intentions in the quest to provide one of the most basic of all human needs: sustenance. We need to improve the way we produce and distribute food. We need to eat, and we need to eat *better*.

A Growing Problem

You might be thinking that I'm playing with the words 'broken' and 'imperfect'. Surely, something that's imperfect implies that it's broken? For a food system to be 'broken', it must once have been intact and this has never been the case.

As a helpful way to illustrate what I'm getting at, let's use an analogy. A (very old) carpenter has been (very slowly) building a wooden dining table. He's been building this table for around 12,000 years. Initially, over the first several millennia, he made great progress. He designed the legs, a tabletop and most of the parts required to put it together. Gradually, as civilisation progressed, he gained access to better and better equipment such as metal tools, which he used to assemble it. A couple of hundred years ago, the technology available to him accelerated, so much so that he eventually had pretty much everything he required at his disposal such as machines that should have enabled him to build the perfect dining table, one that could be used to feed everyone. All he had to do was to implement some of the finishing touches and make sure everyone had access to this table. However, our carpenter noticed that not only was the number of people he needed to feed growing rapidly, but everyone wanted to eat more. Consequently, he decided that his table needed to be both wider and higher. To do this, he had to change the design and attach different materials – not just wood, but modern synthetic materials, too – onto his previously near-perfect design. Unfortunately, in doing this, he damaged many of the original parts. In fact, the table design changed so much that, by the 21st century, it was starting to look very different from what he had originally planned. The dining table was huge: it could seat more people, could provide more food and for longer than he had originally intended, but some simply couldn't reach it and many of those that could hurt themselves on it.

Our carpenter's work remains unfinished, and the dining table may, indeed, be damaged, but it is not broken. Indeed, the fact that more people are now able to sit around it in many ways tells us that it's better than the tiny one from which they previously ate. The table is, however, one that's far from perfect. Nor is it anything like what it could or should be.

The dining table represents our food system. It has to feed an increasing number of people a greater amount of food than what can be provided by the natural resources of Earth's environment, at least in its current guise. Since before the onset of agriculture around

12,000 years ago, the food system has progressively expanded and in many amazing ways it has improved. Human technology has endowed us with the know-how and means to create a system capable of providing sufficient and optimal nutrition for every individual on the planet. Yet the nutritional quality of what we consume is suboptimal for the majority of people and the methods by which we're producing our food are contributing to environmental destruction and causing unnecessary suffering to animals.

Prior to agriculture, when our ancestors survived through hunter-gatherer lifestyles, searching for food was a daily struggle. Hunting and gathering sustenance risked the loss of life: humans were vulnerable to attack by predators, competition from other tribes, injury and, of course, insufficiency. Mortality was probably high and, after a few months, tribes were forced to relocate when the food resources of the vicinity ran out. To overcome this, humans developed agricultural processes as a means of having a reliable supply of food nearby so they didn't have to keep moving settlements. In many respects, these early farming practices worked out well and food was, for the most part, readily available. However, manual farming practices were hugely laborious and a considerable amount of work was required to produce food. Moreover, human anatomical evolution hadn't lent itself to the design of physiques for use this way. As it turned out, manual farming practices required a great deal more effort than their hunter-gatherer ancestors needed to put in. Add to that the risks of drought, flood, pestilence and theft that could instantly destroy the fruits of hard labour: a poor harvest or ruined stored grains meant many members of early settlements starved to death. This led to the requirement to share resources. Sharing your crop yields with a neighbouring settlement who'd been inflicted with a poor harvest acted as a form of insurance. If you were to suffer a similar fate in the future, it was likely that your generosity would be reciprocated. This also meant that one farm needn't be overly reliant on one crop, enabling farms to diversify through trade. In turn, settlements expanded and these grew into cities. As trade advanced, empires emerged and this was followed by the interdependence of different cultures. People had access to wider varieties of foods, further strengthened as agricultural practices were shared, non-native plant species were cultivated and non-indigenous livestock breeds were bred – and then interbred – in different regions. Unfortunately, poor trading practices and clashes of cultures, along with climate and geographical disparities, contributed to conflict. While regions like Europe and the Middle East flourished, those residing in others, like southern Asia and Africa, struggled.

Since the late 19th century, when we saw the first industrial farming methods, the imperfections of our food system have been further amplified. As we successfully solved problems such as cost, availability and sufficiency, others which we'd not previously encountered, such as animal welfare concerns, compromised nutritional quality and environmental pressures, replaced them. New problems required new solutions, so industrial, logistical and technological fixes were developed.

After the emergence of agriculture, humans had to vastly increase the amount of time, work and effort they put into providing themselves, their families and their fellow citizens with sustenance compared to their hunter-gatherer ancestors. Never before had humans had to unduly worry about the effects of drought, flood, pestilence and theft. Certainly, pre-agrarian societies weren't without their issues, but any threat to a source of food merely mandated a redirection of hunting and gathering efforts and starvation could be averted. Moreover, inequality between agricultural systems contributed to reduced trust between groups and increased levels of conflict. By being dependent on only one or two sources of food, early farmers were more likely to suffer when a yield failed. Through living in communities and in close proximity to livestock, infectious diseases that originated in domesticated species quickly spread to humans; their legacies, such as measles and tuberculosis, are still with us today. Some anthropologists have argued that the onset of agriculture may have been the single worst thing ever to have happened to humanity.[3] But is this really the case?

Agriculture's Advantages

In making a comparison between pre-industrial agricultural communities and hunter-gatherers, it's certainly worth questioning whether the invention of farming was a net positive for human wellbeing. Yet, despite the grim image painted above, in a lot of ways agriculture has been a boon for many populations, and we should be cautious of idealising the lives of pre-agrarian societies: life, no doubt, was hard, child mortality was high and a minor accident with a short recovery today could have been a death sentence.[4]

The benefits of farming have led to modern affluence. For instance, farming has removed the necessity for the majority of people to be involved in the production of food. Allowing others to do the farming for them permitted a division of labour and afforded non-farmers the necessary time and energy to adopt other skills, like the production of goods or the provision of services, which could, in turn, be traded with the farmers. Today, less than 25 per cent of the world's population is involved in farming. In industrialised nations it's less than 5 per cent.[5] Agricultural practices and systems of trade have further evolved over millennia, created mostly with the best of intentions in the quest to provide a constant and adequate supply of nourishment for as many people as possible. At the same time, those involved have continued to improve systems of food acquisition and distribution. Modern civilisation would never have happened if it wasn't for agriculture. You would not have been so fortuitous to be able to enjoy the surroundings in which you're currently sitting if humankind hadn't produced the chain of events that our ancestors set in motion around 12 millennia ago.

The picture of food adequacy hasn't always been rosy, of course. Even as recently as the mid-20th century, biologists and sociologists were making doomsday predictions, such as those made in 1968 by Paul R. Ehrlich in his book *The Population Bomb*.[6] Ehrlich predicted that there wouldn't be enough food to feed everyone by the 1980s and that hundreds of millions of people would starve to death. But Ehrlich's dystopian future never materialised. In fact, what transpired was the opposite: an abundance of food coupled with a global epidemic of obesity, non-communicable disease (NCD) and unfathomable quantities of food wasted every year.

Why were these relatively recent catastrophic predictions so wrong? The main reason is human ingenuity. Simply put, we solved the problem of food insufficiency. One notable example is the Nobel Prize-winning plant scientist Norman Borlaug, who pioneered developments in wheat production by creating more resistant, high-yielding varieties. Borlaug's work led to similar innovations with rice and corn and consequently yields in many regions multiplied. Appropriately, Borlaug is widely credited with saving over a billion lives.[7] Farming vast numbers of animals in relatively small areas, and feeding them higher-yielding crops, has, in turn, allowed more affordable quantities of meat to become available. Agricultural developments like these, coupled with improved global logistics, have provided access to adequate protein and calories for the majority of people, even those in the poorer nations burdened with geographical and meteorological hurdles. Human ingenuity has prevented billions of people from starving to death. For these people and their descendants the food system is most certainly not 'broken'.

While the brilliance of the Borlaugs of this world is undeniable and has solved food availability for enormous numbers of people, has the quality of what we eat been compromised? How have modern agricultural practices affected the earth beneath and the sky above?

Declining Nutrition

In 2015, the *New York Times* published a short column called 'A Decline in the Nutritional Value of Crops', citing evidence that demonstrated a significant decline in the nutritional content of the food we eat.[8] The bulk of the article can be summarised as follows: you may now need to eat three oranges to get the same amount of vitamin C that your grandma used to get from one. Before dismissing this as an exaggeration, consider the evidence.

Long-term analysis of the nutrient content of fruits and vegetables in the UK has shown that concentrations of most of the minerals has declined since 1940.[9] Similarly, US studies have highlighted that the levels of protein and key micronutrients have declined.[10] Alarmingly, the trend seems to be global: in 2017, scientists at the National Institute of Nutrition of Hyderabad released data on 151 nutrients collected from foods across India, revealing a significant decrease in nutrient value.[11] The authors of these studies chalk up the declining nutritional content to the preponderance of agricultural practices designed to improve traits like crop size, growth rate and pest resistance. Healthy soil is a prime source of minerals, providing essential support for growing crops. Traditionally, farmers used the crop-rotation technique to preserve soil fertility, which allowed it to 'rest' for a season and recover from the effects of farming. Nowadays, as demand for crops continues to increase, this isn't always viable. Modern, mass-production techniques, the theory goes, are depleting minerals faster than the microorganisms in the soil can replenish them. Essentially, as yields went up, nutrient levels went down.[12] One hypothesis is that increased use of modern fertilisers and pesticides has contributed to the altered composition of micronutrients in the soil which has affected the quality of the food we eat. Combined, these factors mean that over the past few decades new techniques developed to increase efficiency and to make food look more appealing have lowered the nutritional content.

But could there be anything else to it? In the autumn of 1843, agricultural scientist John Bennet Lawes and chemist Joseph Henry Gilbert sowed the first crop of wheat on a field named Broadbalk in Hertfordshire, UK. Every year since, researchers have sown winter wheat on all or some parts of Broadbalk to compare crop yields grown using inorganic fertilisers with those grown using organic or farmyard manure. More than 180 years after the first crop of wheat was sowed, researchers continue to dig up interesting findings from this plot of land. The latest concerns the impact of rising levels of carbon dioxide (CO_2) in the environment on wheat's nutrients. Recent studies have shown that, although the concentration of key minerals like iron, zinc, copper and magnesium remained stable between 1845 and the mid-1960s, they've decreased significantly since.[13]

A 2014 *Nature* paper seems to confirm these findings. The study compared the nutrient content of wheat grown in present-day conditions with wheat grown in an atmosphere with an elevated level of CO_2, similar to what's predicted by 2050. The researchers found that both wheat and rice grown in higher CO_2 environments had lower levels of zinc, iron and protein.[14] The protein decline is particularly important: plants are a key source of protein for many people particularly those in developing regions. The exact reasons for the protein decline are unclear, but growing evidence suggests that elevated CO_2 levels reduce a plant's ability to take up nitrogen, which in turn affects the protein content of food. A 2015 study showed that the effect persisted even after the crops were grown with nitrogen-rich fertilisers, ruling out the possibility that the lower protein content is due to limited access to nitrogen in the soil.[15] Could it be that with a greater amount of CO_2 in the atmosphere, there's less 'room' for other elements like nitrogen? Or is CO_2 in some way interfering with plants' abilities to take up nitrogen?

Further evidence for the declining nutrition of our crops can be found in regenerative agriculture, a method of farming that might also offer a solution. As recently as 2022, scientists have shown that regenerative farming practices enhance the nutritional profiles of both crops and livestock compared to conventional practices.[16] Soil health is likely an under-appreciated influence on nutrient density, positively influencing vitamin, mineral and fatty-acid profiles, and this could have relevance for the prevention of NCD.

Investigating the nutritional content of the food we eat is significant because although getting sufficient protein, fat and carbs is important, it's equally critical that our diet provides adequate amounts of micronutrients, i.e. vitamins and minerals. These substances help keep the body running and protect against both NCDs, such as cancer, diabetes and heart disease, and infectious diseases. A reduced intake might translate to a reduced ability to fight off potential pathogens and protect against chronic disorders. The declining crop nutrient content is a particular concern for those residing in developing regions. It's estimated that more than one-third of people globally are micronutrient malnourished, a figure that includes Western nations,[17] and over half of the world's children under five years old and two-thirds of non-pregnant women of reproductive age are deficient in at least one essential vitamin or mineral.[18] Fruits and vegetables can be key dietary contributors for many micronutrients, so declining nutrient concentrations are concerning.

For most of us, how big a deal is this? When it comes to minerals, the authors of a 2005 paper note that as 'horticultural products in general, and fruits and nuts in particular, are relatively small contributors of minerals to the average UK diet, historical changes in mineral composition are unlikely to be significant in overall dietary terms'.[19] Could it be that, instead of focusing on the pounds, we're fixating on the pennies? Today's fruit and veg consumption is far from ideal in most Western countries. One 2021 study showed that, in the UK, if everyone ate the recommended 400 grams per day fruit and vegetable intake, healthy life expectancy would increase by an average of 7–8 months.[20] Maybe the problem of inadequate nutrition is more fundamental: most of us aren't getting enough fruit and veg in our diets anyway, no matter how they're grown.

The Route to a Perfect Food System

A perfect food system delivers sufficient, affordable, sustainable and nutritious food for every human, reliably ensures adequacy for the future, respects others, both humans and nonhumans, causes minimal damage to the environment, acknowledges cultural diversity, and any resulting adverse consequences are fully mitigated with no unnecessary casualties.

All the problems of the modern system of food provision are resolvable. We already possess the know-how to make vast improvements and the route towards a perfect food system need not be as complicated as some think. We have learnings from history and technological advances that have solved prior food-related problems, we have experts in nutrition, dietetics, food science, agriculture, climatology, sustainability, ecology, logistics, genetics, neuroscience, psychology, politics, economics and philosophy poised and ready to tackle the new challenges. We are motivated and we are altruistic. The journey will be a hard one and, realistically, won't be a quick one. But now is the time for us to map out what we need to do to get there. We have precious little time to waste.

Once we accept that the problems of the food system are primarily due to human behaviour, we can find solutions. Understanding requires a broader perspective, however. I am going to show you how, with a renewed mindset, you can make better choices. My focus will be on what you, as an individual, can do to positively impact on the food system. Although we have a limited ability to affect the behaviour of others, we can look at ourselves and think more carefully about what we choose to put into our mouths.

Personal Diet Contemplation

My first introduction to nutrition was in the late 1970s. My father was 'going on a diet', and I recall him showing my mother a thin ring-bound book with a slender bikini-clad model on the cover. My parents explained that every food contained these things called 'calories' and in order to lose weight you had to have fewer of them. The book, they informed me, was a list of the calorie content of foods and would help Dad monitor how much he ate. That was the point when nutrition science had entered my life.

But it was through my mother that my interest in food and nutrition really accelerated. When I was eight years old, she was diagnosed with breast cancer. After undergoing the best treatments available in the 1980s, she felt that she needed to do more to help prevent the disease from spreading, to support her health and improve her quality of life. This involved her making a bunch of lifestyle changes that included chugging daily vegetable juices, taking a load of supplements and switching to a vegetarian diet. Although she didn't force vegetarianism on me, I was fed a number of what, to me, weren't 'normal' meals – bear in mind that, in rural England back in the eighties, being a vegetarian was still fairly niche. I was also privy to a lot of nutrition and health-related conversation around the dinner table. Although some of the strategies Mum tried are the kind of thing I now find highly questionable, her diet did involve many sensible food choices. Indeed, her approach may well have been beneficial, as it wasn't until 11 years later when she succumbed to the disease: I'm confident that her lifestyle changes helped her live, for the most part, in reasonable health for these additional years. And fortunately, she got to see her son go off to university to study nutrition. Mum's interest in nutrition was definitely a key factor in kick-starting this now 30-plus-year-long vocation of mine.

The other key element that propelled me into learning more about diet and nutrition was my obsession with bodybuilding. When I first stepped into a gym, I was a skinny, 125-pound wet-through 16-year-old, but I instantly caught the weight-training bug. From that point, my life had a single purpose: to pack on as much muscle as possible as soon as possible. Realising that good nutrition would be key if I wanted to look anything remotely like Arnold Schwarzenegger, I read book after book on anything to do with nutrition for muscle growth. After necking an array of disgusting supplements and hungry to learn more – greater knowledge equals bigger muscles, right? – I decided that I should absorb as much information as possible about diet and nutrition. So I applied to study nutrition and dietetics at university.

Even before I'd packed my bags to go off to university, I knew that I wanted my career to involve diet and nutrition. So, four years later, after graduating, I did what most UK dietetics graduates did: I got a job as a clinical dietitian in the National Health Service (NHS). At that time, I didn't realise what an incredibly useful experience this was going to be. During my seven years in the profession, I supported thousands of patients and got to see a range of medical issues that required dietary assistance. My role included prescribing specialised tube-feeding formulas to patients who were unable to swallow, were burdened with cachectic muscle wasting or who simply weren't eating a sufficient amount of food to support their recovery. I also advised patients who needed help with weight control, diabetes, food intolerances, eating disorders and other dietary interventions. The experience I gained in the NHS bestowed learnings that would later be of huge benefit to my career. For example, I got to see that providing basic nutrition to critically ill patients could translate into rapid recovery, I learned the extent of the effect that income disparity has on the quality of people's diets, and from working with people with obesity I came to realise that being 'overweight' is way more complex than simply being 'lazy'.

After my stint in the NHS, I diverted my career down a different path. Fulfilling my teenage passion, I started my own business in the gym and fitness industry. This gave me the opportunity to support athletes and sportspeople of all ilk with their dietary needs, including bodybuilders, strongmen, fighters and recreational gym-goers. My clients ranged from people new to the gym to those who competed professionally at international level. This 16-year period of my career gave me a front-row seat to witness the sheer commitment mandated by elite sportspeople in order for them to excel at their event. Equally, I got to see how, for novice trainers, small changes in their nutrition could translate into phenomenal gains. My nutrition work during this period wasn't limited to advising sports and fitness enthusiasts; I also advised clients to help them control their weight, improve their cholesterol profile, manage allergies and more. Among other projects, I wrote articles for magazines, newspapers and websites, gave lectures and seminars, and supported care homes and institutions with meal planning for their residents.

The next deviation in my career came in 2015, when I co-founded Huel, one of the fastest-growing food and nutrition brands in the UK. Huel produces nutritionally complete, convenient and sustainable food. I devised the recipes for all the initial Huel products, and even today – more than nine years on – the nutritional profile of every product in our range is approved by me. My role has included selecting the most suitable ingredients, educating and advising potential customers, suppliers, colleagues and potential investors, being involved with research trials, working with healthcare professionals and taking the lead in all matters concerned with nutrition and sustainability in the business. Being co-founder of Huel has taught me so much. Crucially, I've had to consider the nutritional needs of populations: after all, millions of people rely on the convenience of Huel products as a key contributor to their daily nutrition. This has been both a huge responsibility and a privilege. I've been forced to open my eyes to the many other aspects of nutrition that I'd overlooked earlier in my career, such as the effects of our food system on climate change and food waste, and to consider the ethics of food production and how it affects human rights and animal welfare.

Contemplative Nutrition

Naturally, as a nutritionist, I've always put a great deal of thought into my own dietary choices. However, these four stages in my career – student, clinical dietitian, bodybuilding and fitness nutritionist and sustainable food business leader – bring me to where I am now: a nutrition communicator. I've discovered that there's a whole lot more to educating on nutrition than simply telling people what nutrients and foods they should eat. Through reflective consideration, I've learned to view nutrition through a different lens. No longer would I eat merely to support my own health and to aid my performance in the gym. When it comes to making my own food choices, I should take other crucial issues into consideration. This led me in 2019 to change my diet and to adopt a way of eating that I've since named 'contemplative nutrition'.

What I will propose is a dietary strategy that starts with you. Your food choice will influence, first and foremost, your own health and wellbeing, but it also impacts on other humans, nonhumans and the environment. What we choose to nourish ourselves with in the 21st century involves us having to pause and consider what we eat. Contemplative eating is a dietary philosophy that provides a solution to the problems of the modern food system through five key focus points – or 'pillars' – of contemplative nutrition. Contemplative eating will help you choose foods that will improve your health and wellbeing while simultaneously addressing environmental and animal welfare concerns. Ultimately, we want to eat good food that we enjoy: we don't want to have to contemplate what to eat *too* much!

You are about to learn what contemplative eating involves and why you'll feel better from eating this way. But more than this: you'll understand why contemplative nutrition could be a way of eating for everyone. Contemplative eating is not prescriptive, it's not a fad, nor is it a 'diet' in the rigid sense that people think of when they hear the word 'diet'. Crucially, to eat contemplatively need not be difficult: the food choices you make may not be all that different to how you're currently eating. As you'll soon see, all that's required is for you to think – just a little – about what you choose to eat. It's a way of eating that will help you navigate the highly nuanced world of nutrition by embracing different perspectives.

Contemplative nutrition is a mindset.

Chapter 2

Our Drive to Eat

Fifteen thousand years ago, our ancestors weren't paying attention to the protein content of their food. Nor were they thinking, 'Am I getting enough vitamin D?', even during the Ice Age, when there was less exposure to sunlight. People didn't think, 'I haven't had enough omega-3s this week' in fishing communities or otherwise. And I doubt that their hunting arsenal included the latest trending calorie-counting app. Although the acquisition of sustenance was likely considerably hard, Paleolithic humans had few worries when it came to thinking about what they fancied for their next meal. Back then, sufficiency was the primary consideration: humans simply ate what the tribe hunted, gathered and shared.

I also doubt that our great-grandparents gave much thought about a meal's carbon emissions, and, when it came to farming, what they deemed ethically acceptable was probably very different to how people feel now. The concept of having the ability to choose what to eat is a relatively new phenomenon. Compared to in the past, the vast majority of humans today are fortunate to be presented with a considerable choice in respect of what foods are available, especially in the West. A lot is known about the nutritional value of our food and, more recently still, many of us now consider other important issues like the ethics of food production and sustainability when it comes to our diets.

Earlier I stated that it's the individual who has the final say in what they choose to put in their mouth, at least in modern Western societies. But how much of your food choice is really down to your own free will? To comprehend what it means to eat contemplatively and to change the way we eat, we should first understand what compels us to choose the foods we do. The anthropologist Marvin Harris noted that failing to understand the practical basis for food preferences and aversions seriously impedes attempts to eat better and can 'lead not only to ineffective remedies but to dangerous ones'.[1]

Contemplative eating is about why we should and how we can make better food choices. To understand what this involves, we must first understand why we choose to eat the foods we do. This isn't always obvious, especially in modern developed regions where we take sustenance for granted. With food a-plenty, few are constrained by lack of availability. Where people have access to a wide range of high-calorie convenience foods there's a greater risk of obesity and other non-communicable diseases (NCDs). For example, a 2014 study demonstrated that exposure to takeaway food outlets was associated with higher consumption of foods purchased from these outlets and greater body weight and obesity risk.[2] In the age of fast food, the influence of convenience is no surprise. The global fast-food market was worth nearly $862 billion in 2020.[3] If you're tired or busy, preparing a meal may be low on your list of priorities. As well as what's available, how much food costs will hugely influence your choice. In the 21st century, a greater number of people have the financial means to buy more food than ever, but their choices will still be restricted by their budget. Low-income households are less likely to buy as much fruit and veg as those with a higher disposable income.[4]

But why do we eat? The obvious response to this question, of course, is *to provide the energy and nutrients necessary to sustain life and flourish.* But this answer fails to fully address the question. Implicit in it is one more pertinent to why we choose the foods we do: what *drives* us to eat?

Biology's Overwhelming Grip

Have you ever considered how much control you truly have over your conscious experience of eating? When we start to explore our most basic motivations to eat, we uncover the overwhelming grip that our biologies have over our drive for food. When you go to the store to buy groceries, you might feel like you're in control, but it's far from straightforward. After all, if our dietary choices were solely dictated by our active consideration of what and how much to eat, why do we buy foods that we know are bad for our health?

Our food choices are strongly influenced by the attributes linked to the pleasure we derive from what we eat: the taste, flavour, aroma, mouthfeel and presentation of food. Food is frequently a topic of conversation: we're obsessed with it. Chefs spend time creating flavoursome dishes adorned with visual cues, and presentation is a crucial attribute for success in cookery competitions. Retail establishments attempt to manipulate the organoleptic pulls of passers-by to reel them in, like the bakeries that purposely encourage the enticing aroma of their freshly baked products to drift from their doors.

Understanding which factors are at play in driving our motivation to eat has long been an area of fascination for nutrition experts and brain enthusiasts alike. Despite this, the full interplay of the signals that drive our motivation to eat remains frustratingly unknown. What we do know, however, is that from a neuroscience perspective, the motivation to eat is no different from the motivation to engage in any other human behaviour. The regulation of food intake requires a fine balance between excitatory and inhibitory processes. The excitatory processes arise when your body is running low on energy and you feel hungry. These incoming cues may include heading to your fridge, nipping to a store or reaching for your phone to order a delivery. Goal-directed behaviour also hinges on our ability to ignore potential distractors. These inhibitory processes may be triggered by the physical and mental effort required to get food. For example, after you've eaten, signals are released in your body that decrease your drive to eat. Here, the inhibitory signals are more powerful than the excitatory ones with the end result of tipping the scale towards a reduced motivation to eat. Human behaviours arise from a series of competing signals and for any action to be initiated – whether it be eating or anything else – excitatory signals must dominate. Crucially, most of these processes operate completely beyond our conscious awareness: you decide to eat because a constellation of intrinsic and extrinsic cues have aligned to drive your motivation to eat.

A glaring example of the hidden forces that motivate us to eat comes from food cravings. Food cravings are an adaptation for us to seek out energy-dense nutrition, and are what motivated our ancestors to pursue hard-to-come-by sustenance, such as providing the impulse to climb up a tree to grab some delicious-looking fruit. Although myriad mechanisms have been described to trigger these goal-oriented behaviours, a key chemical linked to our motivation to seek out hedonic food is dopamine. This is because perhaps one of the best-described mechanisms linking dopamine to human behaviour has to do with its role in learning about rewards. If you're a chocoholic, there's every chance that the indulgent feeling of chocolate melting in your mouth will stimulate an intense feeling of pleasure, which promptly translates into the release of copious amounts of dopamine in your brain's 'reward system', the term that scientists use to describe the part of the brain that lights up on scans when you're enjoying something.[5] Neuroscientists believe that dopamine's dash in this circuit is the strategy that evolution has developed to reward us for engaging in life-sustaining behaviour, although it's also been linked to a number of addictive behaviours like gambling, sex and drug use.[6]

Dopamine *reinforces* highly rewarding behaviour but the extent that it *influences* it is unclear. Many scientists have shown that the concentration of this neurotransmitter* can have far-reaching effects on our tendency to feel pleasure, but when it comes to understanding the subtle science that drives our behaviour things get a bit messy. Partly this is because dopamine has been traditionally associated with reward processing – how good or bad an outcome is after taking a decision – but not conscious experience.[7] In other words, juicing up dopamine levels in the right brain areas can help you to pursue valuable rewards – the sight of molten cheese on your favourite pizza will likely spark a surge in dopamine levels, causing you to want to eat it. However, given that some aspects of reward processing require conscious decision-making – like weighing up the perceived enjoyment from munching the pizza with the regret you'll likely feel later – it remains unclear if dopamine may also be active in regulating our conscious experience of reward pursuit. Regardless of the exact role that this chemical plays in regulating cravings, addiction and, ultimately, food intake, it is important to note that hedonic processes alone are sometimes insufficient to control your appetite. Indeed, there is broad consensus among scientists that hunger and appetite are in fact the by-product of homeostatic processes which often interact with hedonic processes to generate a powerful drive to eat.[8]

Hunger Hormones

As well as being driven by cravings, when our brains want to know how much we should eat, two other signals come into play: appetite and satiety. Appetite is the desire to consume food; it's how our bodies

---

* A neurotransmitter is a chemical messenger that transmits a signal from one neuron to another.

help to regulate energy intake. Satiety is the feeling we get when we've eaten enough. Both are under the control of several hormones.

Around 30 years ago, endocrinologists discovered and named a hormone that's now become so fundamental to any understanding of obesity that it's mentioned in just about every weight loss magazine, website and book. Leptin can be thought of as our weight's master control chemical.[9] It's a controller of our appetite and it's released from adipocytes (fat cells). Our bodies don't regulate leptin release; rather, the more fat you have, the more leptin is released. It signals the hypothalamus – a region of the brain that's involved with weight control – where it acts on both hunger and certain metabolic reactions to control how much energy we take in and how much we burn. The role of the leptin–hypothalamus feedback system is to keep our bodies at a healthy weight with a sufficient amount of energy stored as insurance in case of an impending short-term famine. As the volume of fat in our bodies increases, the amount of leptin in the blood rises and our hypothalamus tells us that our energy stores are adequate and no more food is needed. Satiety and metabolism both increase: we feel full and burn energy faster. Conversely, if the amount of fat decreases – say during famine or illness – the hypothalamus detects a lower level of leptin in the blood and acts to prevent further loss of energy by reducing our resting metabolic rate and increasing our appetite in an effort to preserve valuable fat stores. When food's back on the menu, we eat more, body fat increases, leptin goes up and weight is regained. The explanation here is, of course, overly simplistic with numerous other factors coming into play. The overriding influence of leptin is highlighted in insulin resistant individuals (insulin is the key hormone in blood glucose regulation) and in those whose bodies have a high degree of inflammation, possibly due to stress or illness. In these folk, leptin's effects are weakened – known as leptin resistance – and they find it both harder to feel full after eating and to lose body fat.[10]

Discovered a few years after leptin was another hormone: ghrelin, a chemical produced in the cells of the upper stomach.[11] Ghrelin plays a role in the stimulation of hunger, in particular in response to food deprivation.[12] Under normal conditions, an empty stomach prompts the release of ghrelin into the bloodstream, and when circulating levels are high enough, the hormone reaches the brain, where it activates cells in the hypothalamus.[13] Once we start eating, ghrelin levels gradually decline as the stomach starts to expand and we start to feel less hungry. Ghrelin levels have been correlated with how ravenous obese people who are dieting feel: it's one of the reasons why dieters struggle to control their appetite when trying to eat less.[14] Thinking about food preoccupies their thoughts to such an extent that it's difficult to concentrate on other tasks. The more thoughts of food occupy our minds, the more our senses become tuned to organoleptic cues and the more we want to eat. Ghrelin is considered the key hunger hormone: it tells us to go and seek out food and rewards us for doing so by working with dopamine by increasing the pleasure we derive from eating.

Other hormones that are involved in how full we feel include peptide-YY (PYY), cholecystokinin (CCK) and glucagon-like peptide-1 (GLP-1), all released by the cells of the upper small intestine when they detect the presence of semi-digested food. PYY lets our brains know when we've eaten and the urgency to acquire sustenance has subsided. PYY acts on the hypothalamus and increases short-term satiety. It's responsible for the feeling you get when you've just eaten and you begin to lose the desire to seek out more food.[15] CCK is involved both in digestion and hunger suppression. Released rapidly in response to food intake, especially when fat is detected, it suppresses appetite by inhibiting stomach emptying and sending signals to the brain. CCK curbs hunger to prevent us from overloading our bodies with too much fat in one sitting, which would be hard to digest.[16]

GLP-1's effects are even more pronounced. It helps to quieten 'food noise'. Food noise is defined as the 'heightened and/or persistent manifestations of food cue reactivity, often leading to food-related intrusive thoughts and maladaptive eating behaviours'.[17] When people have a lot of food noise, food is constantly on their minds: subconsciously, their brain is working on overdrive to find food to keep them alive. As well as signalling the hypothalamus to tell us to put our forks down, GLP-1 also plays another key role in the regulation of energy balance: it acts on the pancreas to make the effects of insulin more effective.[18] The level of insulin goes up after we've eaten a meal and a stable blood glucose level is crucial for a range of healthy metabolic functions. So, although we might start to feel a little full after that second wedge of cake, insulin needs all the help it can recruit to stop things from getting out of control following the massive influx of sugar. GLP-1 reminds us that maybe we shouldn't be having that third slice of chocolate gateau that's staring us in the face, while it lends a helping hand to insulin in controlling our blood glucose.

GLP-1 has taken centre stage over the last few years due to the weight-loss drug semaglutide. Semaglutide, commonly known by the brand names Wegovy and Ozempic, is a member of a class of drugs known as 'GLP-1 agonists' which mimic the actions of GLP-1. Originally developed as a drug to help diabetes, semaglutide can now be prescribed for weight loss in several countries,[19] and it's very effective: users report that they lose their enthusiasm for food and start to feel full very quickly after eating just a small amount. Since 2021, semaglutide's popularity has skyrocketed after Hollywood stars in their droves have shouted about its effectiveness on their social media accounts. Semaglutide is now being used by hundreds of thousands of people as an effective means of weight control.[20] It turns down food noise.

Organoleptic Cues

The organoleptic cues – the smell, taste, texture, temperature and sight of food – are crucial influences on our food choice. The flavour of a food refers to its perceptual impression and is determined by the smell, taste, temperature and texture of food, with smell contributing the largest proportion. Smell detection involves the olfactory system, principally our noses. Our sense of smell is a highly complex but poorly understood sense, mainly because, as smell is a subjective phenomenon, there's limited value from performing experiments with animals. Moreover, humans have a poorly developed olfactory detection system compared to other mammals. Dogs, for instance, have around 300 million olfactory receptors in their noses; humans a mere six million.[21] We do, however, have at least 100 primary sensations of smell[22] and these interact in multiple combinations and intensities, inferring that the number of food-derived smells is unlimited. This means that the number of flavours our brains have the potential to detect is unbounded. Although, compared to other mammals, humans are relatively poorly tuned when it comes to the detection of smells, the fact that smell has a major influence on a food's flavour makes our olfactory sense crucial to the value we attribute to our foods.

Taste is detected via taste receptor cells situated within our taste buds. Each taste bud contains around 100 taste receptor cells, and there are between 3,000 and 10,000 taste buds primarily located on the tongue, with others situated on the roof, sides and rear of the mouth and the throat. There are five distinct primary taste modalities: sweet, sour, bitter, salty and umami (savoury), and a particular food will stimulate different combinations of the five modalities.[23] The taste buds responsible for pungency are involved in the detection of spicy foods, and our gusto-olfactory system also detects 'coolness' (such as menthol) and oleogustus (fattiness),[24] which may vary in their degree of subjective perception, an effect that may be culture-dependent, indicating that the degree to which they have an effect can be learned.[25] Flavour is also determined by the temperature and texture of food involving our palates' thermoreceptors and mechan-oreceptors, respectively.

The unbounded number of oral-nasal organoleptic cues sent to our brains culminates in our hypothalamuses being showered with a plate-shattering number of sensations with every mouthful, influencing the effects of dopamine and other chemicals in the tightening of our biology's grip. This serves to remind us that our conscious control of what we put in our mouths may not be as great as we think. And that's not all: often the first sensory encounter our brains have with food is via our eyes. The influence that visual cues have on our food choice is, unsurprisingly, poorly studied, although no successful chef will need reminding of the critical importance of presentation to culinary perfection. In fact, a number of visual factors associated with what we choose to eat have been identified including proximity, visibility, colour, variety, portion size, height, shape, volume and surface area of food.[26] Now that you're aware of the numerous influences interacting in myriad different ways that affect what you put in your mouth, can you truly assert that the foods that you select are the sole consequence of your own volition? Can you honestly conclude that you're the conscious author of the script behind what you choose to pick up with your fork?

This is worth reflecting on for a moment as it's key to how we humans view our food and is important to keep front of mind as we consider each of the contemplative nutrition pillars.

Chapter 3

Contemplating Nutrition Misinformation

I'm going to open this chapter by saying something that risks my credibility, may disappoint you and might even make you want to slam this book shut: *I have 30 years' experience working in the field of nutrition and I don't know what an optimal diet looks like.* Worse: *I can't even come close!* If me saying this makes you feel that I lack validity, I'm more than okay with that. I just don't want you to buy into the nutrition bullshit* that's thrown around. It's my duty, as an experienced professional, to help you steer away from the dietary misinformation you're exposed to every day from the food industry, marketers, policymakers, social media influencers, and even many scientists and nutritionists. I'll sleep soundly at night if this affects how people feel about my credibility because *no one* knows what an optimal diet looks like. This is for one simple reason: there isn't one. What does 'optimal diet' mean anyway? Optimal in terms of how many years it adds to life? Optimal for lowering non-communicable disease risk? Optimal for losing weight? Optimal for what age, sex or body weight?

I do, however, know what a *poor* diet looks like and I bet you do, too. The scope of what can be described as 'poor quality diets', however, is vast. It ranges from the diets of people in poorer nations

---

* I'm not swearing here. 'Bullshit' is speech that's intended to persuade without regard for the truth.

who have limited access to adequate nutrition to those of us in the developed world with such a wide array of foods to choose from that many of us are consuming too much.

Now that we've established that we know what poor diets are despite not knowing what an optimal diet is, we should ascertain what suitably nutritious diets are. This is where things get interesting. There have been multitudes of dietary strategies published in numerous books and websites over many decades. Each regimen has its proponents lined up ready to defend their nutritional ideology. I've named this 'dietology' and it refers to the dogmatic commitment to a nutrition-related belief that's not based on a critical appraisal of the available evidence. Crucially, even when presented with otherwise convincing evidence to the contrary, the one holding the belief may even double down on their claims further. Dietology camps include the 'low carbers', the 'low fatters', the 'ketos', the 'paleos', the 'carnivores', the 'carb cyclers', the 'juicers' … I could go on. And 'ideology' is the right term for this mindset: just look at the number of social media accounts that include mention of the diet 'tribe' to which they ascribe their allegiance – like 'keto.james' or 'melanie-the-carnivore'* – proudly worn like a badge of honour. Moreover, what makes this issue particularly hard to navigate is that often each camp is able to present a great deal of seemingly credible peer-reviewed evidence to back up its claims. How can we be expected to discern the truth? At the very least, we want to know if there's a preferable dietary strategy that will support a long, healthy, disease-free life, while enjoying our time being active to the maximum of our capabilities.

---

* The names are made up to illustrate my point (using my own and my wife's first names).

Dietological Incentives

'Avoid oats; they cause inflammation!' 'Fruit is just sugar; it raises insulin!' 'Avoid fish or you'll poison your body with mercury!' 'Cruciferous veg lowers your testosterone!' 'It's got more than four ingredients, so it's ultra-processed!' Day after day, we're bombarded with confusing and conflicting information from social media influencers, demanding that we follow their nutrition advice. Spend a few minutes scrolling through social media and your feed will show multiple fitness influencers blurting 'facts' about some food or another.

Social media is an unavoidable part of the 21st century. In 2023, nearly 4.89 billion people worldwide were using social media, a figure almost double that of 2017.[1] What's more, users are spending more and more time on their favourite channels. With people spending more time online, social media networks have emerged as robust content marketing and branding tools, paving the way for hordes of increasingly influential 'digital creators'. These content producers are influencing our behaviour: a study from the US showed that more than one in three adults routinely relies on social media influencers to buy their products.[2] From which books you read to dating habits, online gurus have rapidly slipped into intimate and personal aspects of our lives. But when it comes to how online food influencers are affecting our nutrition beliefs, we have somewhat of a different threat.

Personal endorsements used to be about harnessing someone's celebrity status or utilising the insights of experts. However, the recent proliferation of online food influencers who have mistaken posting videos with having a career in the food industry suggests that the educational food environment has morphed into something completely new. A brief scroll through Instagram and TikTok reveals an ever-increasing number of content makers sharing their wisdom and diets that, they claim, will improve the health of their followers.

To those outside nutrition and dietetic circles, the scale of the phenomenon might not seem immediately obvious. Indeed, a quick glance at studies on the impact of influencers on dietary choices reveals that the academic literature is alarmingly scant. Most studies published to date show that influencers lead to 'greater purchase intentions' due to participants identifying, relating to and trusting influencers, in some cases more than actual celebrities.[3] In a 2022 study, one in five adults admitted to regularly following food influencers to identify their munchy of choice.[4] Perhaps the most convincing evidence of the sway that influencers have over people's food choices comes from a study published in 2019, in which 176 children were split into three groups. One group of kids was shown an influencer with unhealthy snacks, another was shown an influencer with healthy foods and the last group was shown an influencer with no food products. Children in the group that saw the influencer with unhealthy foods consumed 32 per cent more calories from unhealthy snacks and 26 per cent more total calories than the kids in the group who viewed the vlogger with non-food products.[5]

Why does it matter? Food habits have, after all, always been influenced by others. Social norms have underpinned the practice of eating for millennia and food behaviours have been integral to all human cultures since, well, there's been culture. And outsourcing dietary advice is nothing new: magazines, newspapers, books and TV have been pushing fad diets for decades. Food influencers might save customers time and perhaps even money as they buy stuff they actually like. Or do they?

Starting with social norms, the recent proliferation of all-knowing food prophets dishing out improbable hashtags suggests that today's content could hardly be compared to yesterday's word-of-mouth recommendations. Neuroscientists warn that, anthropologically speaking, digital media and food endorsements are a very recent development in human history and our brain hasn't quite caught up with this.[6] Powerful visuals of palatable food have an outsized impact on our imagination because our brains are engineered to identify resources in an environment that's both scarce and potentially dangerous. These adaptations give a lot of room to content creators to capture our attention with potentially harmful consequences. One 2019 UK-based study found that social media influencers with cult-like followings gave incorrect health advice 90 per cent of the time,[7] and a 2024 study that examined nutrition-related content on influential Australian accounts found that nearly half the posts contained inaccuracies.[8] The study also noted that posts authored by nutritionists and dietitians were found to be of higher quality, but that better engagement was associated with worse quality. It's much easier to make catchy videos that go viral if you don't get bogged down with all the hard work involved in cross-checking your content.

What this means is that in a hyper-connected world, food information is being chewed and digested more quickly than ever. Although the link with social media might not seem immediately obvious, as early as 2014 neuroscientists concluded that Facebook triggers the same impulsive part of the brain as gambling and substance abuse.[9] This brings us to our second point: money. According to Reuters, in the 12 months prior to September 2023, Americans spent $71 billion on online 'impulse' purchases.[10] When it comes to food shopping specifically, marketing agencies offer a glimpse into the influencers' reach. One marketing consultancy website proudly boasts that 'compared to the average food or beverage brand ... influencers harness 5 times better results'.[11] The recipe is simple: the more likes or comments a post generates, the greater the return on investment, with some companies making as much as $6.50 for every dollar they invest in influencer campaigns.[12] The influencer market has a dirty secret: it benefits from a vast and potentially dangerous impact on people's food choices. With social media an integral part of our daily life, it's too late to reverse this trend, and, given the overwhelming number of influencers posting their life online, a mass extinction of camera-holding, frantically gesticulating sages is unlikely. What's more, as AI technologies become ever more sophisticated, it's entirely possible that in the not-so-distant future, apps like Instagram will be serving a diet of clips purposefully made by AI that is acutely in tune with a user's viewing behaviour.

What to eat is a contentious issue. Eating is something we all do, so it's understandable that everyone has an opinion when it comes to food. But this lack of accountability across the board means we've created the ideal environment for charlatans. In a world overflowing with nutrition information, all dietary advice online should be approached with caution.

Anti-Science and Fearmongering

With so much conflicting information leaching into our brains, it can be hard to extract the useful from the trash. There have been snake-oil salesmen, of course, since humans have been trading with each other. But now we find ourselves in a world with social media at its core, with the bulk of health and fitness influencers pushing fake information to suit their own agendas. Naturally, one of the main motivators for social media ultracrepidarians is probably financial gain. They produce clickbait content to attract potential customers into purchasing their goods or services. But I'm not so sure this is always what primarily drives them to keep posting content. Online incentives needn't be solely monetary: popularity, gaining followers and ideologies are equally persuasive motivators. While it might, indeed, be true that they make a living from their content, in many cases these faux influencers seem to display a narcissistic disregard for what's true and are driven by gaining admiration. They have a Messiah complex.

Worse still, I can't help but feel that it's become cool in certain camps to disagree with any mainstream view. There seems to be an anti-conventional-science rhetoric floating around, which is especially rife when it comes to health information and seems to have worsened since the pandemic. From the supposed emasculating effects of soy products[13] to bizarre theories suggesting that governments will force us to eat insects in the name of stopping climate change,[14] what in the past might have been dismissed as a fringe phenomenon is now taking centre stage. Here's a thought: could it be that nutrition science is merely another victim of anti-establishment conspiratorial ideologies?

When untrained individuals cite research, they are liable to interpret the findings incorrectly. Of greater concern is when a study's findings are deliberately misrepresented in order to push a particular narrative. Of course, you don't need to be an academic to learn the necessary skills to be able to interpret a scientific paper, but a worryingly large number of people who adopt the persona of a nutrition expert lack the competency to accurately communicate research findings. Influencers, sometimes with hundreds of thousands of followers, disingenuously drive their unqualified claims for their own dogmatic motivations. This misinformation ranges from echoing something they've heard from another pseudo 'expert' to outright lies and deliberate deception. I don't doubt that, in some cases, influencers come with good intentions and truly believe their ideas are of genuine benefit to the health of their followers. However, lacking substantiation, any assertion is likely ideological and indicates an alignment with a particular dietary camp. For some, their very identity is attached to their nutrition beliefs. An honest actor who truly wishes to disseminate only reliable information should seek to be mindful of their own failure to understand the facts behind the claims they make, or they should refrain from commenting at all. A nutrition message that lacks the required level of rigour behind it could be hazardous to people's health.

You might be wondering how, if these quacks are doling out misinformation, they manage to conjure up huge followings. False prophets endow a sense of control: *change what you eat and your life will be better.* By preaching easy-to-understand sound bites to the masses, their disciples feel more empowered in this otherwise scary, complex, out-of-control world. Heed their advice and *all your ailments will vanish and you'll enjoy a long life, free from disease.* They're your saviour. This black-and-white thinking makes them extremely compelling communicators and goes some way to explain their huge fan base. In the alternative and social media space, they seem to have been able to find an audience that wants to be pandered to in such a way that it doesn't care that what their demigod is telling them is truth-tracking. Nor do their disciples even attempt to track the truth themselves.

However, let's not ignore how these online personas develop huge followings in the first place: as the world grows increasingly more digital, should the blame lie solely with the influencers? Shouldn't the platforms themselves be taking some of the responsibility? After all, algorithms favour posts that have a greater number of likes, comments and shares, and shorter videos get better engagement because viewers are more likely to watch them to the end, further favouring the algorithm. And without fans, there would be no audience for these sham sages to preach to. It's been shown that lies and misinformation spread both faster and farther than factual information,[15] and that negative posts spread more rapidly than positive ones.[16] Humans, it seems, are wired up to chase news that's more novel and attention-grabbing than news that's positive or true. To take advantage of this glitch in the human psyche, one particularly pernicious tactic some content makers use is to demonise particular foods, and not just processed foods or branded products, but fruit, cereals, milk and other items that have been providing humans with precious sustenance for many thousands of years. By vilifying foods that aren't in line with their narrative, they seek to scare people into shunning foods that, they assert, 'aren't meant for the human body'. By using phrases like 'spikes blood sugar', 'increases insulin', 'causes inflammation' and other medical-sounding terms, they fearmonger their audience, yet they provide no explanation of the physiological mechanisms behind their claims nor the adverse health consequences.

While none of this sounds great, so what? After all, despite villainising some items, a lot of pseudo-influencers' content does include healthy, nutrient-rich foods, and I've seen many images of delicious-looking recipes amid otherwise questionable content. Moreover, their content is plastered with testimonials from fans claiming that they feel fantastic after following the advice of their favourite guru. So, if people are appreciating the advice, why does it matter if a certain communication style is a little unorthodox? What is it that's so concerning about the information these zealots preach?

Rejecting foods like fruit, veg, cereals and even some processed foods could result in people missing out on valuable nutrients like fibre, essential fats and key micronutrients. Crucially, the repercussions of dangerous nutrition content aren't limited to the physiological. There are equally concerning environmental and psycho-social impacts, too. Not only are we living in a time of anthropomorphic climate change, ecological destruction and demographic growth, many families continually endure day-to-day financial pressures, ever-demanding kids and all the other stresses of modern life. If the majority of people were to adopt the advice doled out by some influencers – such as by those who advocate meat-heavy diets – the food system, in its current guise, simply wouldn't cope. Moreover, contrary to popular belief, as we'll see later, obesity isn't a choice: maintaining a healthy weight is disproportionately harder for some people than others. As well as this, there are foods shunned by influencers that are linked to particular cultures; villainising them risks alienating people based on their cultural preferences. Consequently, when considering all these concerns, much of the advice from unqualified influencers is, frankly, unrealistic. What's preached is a dietology marketed as a one-size-fits-all when, in reality, it's a one-size-fits-none approach. It seems that for many online charlatans, preaching their dietology is what's truly important to them, rather than actually caring about the wellbeing of their audience. What are the benefits of trying to make people feel like they must buy a particular food or behave in a certain way when they're already experiencing multiple burdens in their daily lives? With many people impacted by socio-economic woes and financial tensions, influencers who fearmonger affordable foods while promoting expensive alternatives reek of unrealised privilege.

Then there's eating disorders, which are among the deadliest mental illnesses, with cases continuing to rise. As well as endowing humanity with improved access to sustenance and a wider range of foods from which to choose, globalisation has led to an increased number of people with a poor relationship with food and a negative self-image. According to a systematic review, the global prevalence of eating disorders more than doubled from 3.5 per cent for the period between 2000 and 2006 to 7.8 per cent between 2013 and 2018.[17] And not only are more people struggling, but kids with eating disorders are starting younger. The typical age of onset since the 1970s has been between 15 and 19 years old, but over the last 20 years a greater number of kids under 14 have been experiencing disordered eating.[18] The reasons for this are complex, but platforms like Instagram and TikTok are partly to blame: not only do they encourage fitness influencers to post selfies, compelling people to aspire to unrealistic goals, they permit pseudo-gurus to dictate what their audiences should and shouldn't be eating.

But the main issue with fearmongering foods is actually much more fundamental. It's pointless. Let's say you have a really terrible diet and you eat one nutrient-packed bean salad. Guess what happens to your health? Literally nothing! But for some reason, the fearmongers act like eating just one delicious, high-sugar food is going to harm your health. By the same logic: if you have a super-nutritious diet and you eat one large chocolate bar, what will happen to your health? Again, literally nothing! What matters for your health is not individual foods, but overall dietary patterns and life-style.

Should We Trust the Experts?

A 2018 BBC documentary called *The Truth About Carbs* included a segment that's been uploaded on the BBC YouTube channel called 'The shocking amount of sugar hiding in your food'.[19] In it, the presenter, along with a Registered Dietitian – i.e. a member of the profession that we should be able to trust when it comes to quality nutrition advice – asks volunteers to play 'blood sugar bingo' by estimating the amount of sugar (measured in number of sugar cubes) that certain foods break down to. The foods include a muffin, bagel, strawberries and white rice. The dietitian reveals that the rice breaks down to considerably more sugar than the muffin and bagel (20 sugar-cube equivalents vs 10 and 11, respectively). This is incredibly misleading: our bodies break down carbohydrate in different foods at different rates. The inference was that rice 'contains' a lot of sugar and should therefore be avoided, when, in actuality, white rice doesn't contain sugar. When supposedly reliable sources like the BBC and Registered Dietitians are spewing out highly misleading and scaremongering information like this, how are we to know what nutrition advice is trustworthy?

Nutrition science is rife with wishy-washy answers to specific queries. How much protein do I need to grow 18-inch biceps? What calorie deficit do I need to lose 20 pounds? Will eating saturated fats give me a heart attack? Humans seek convenient, robust solutions to hugely complex issues, and nutrition science provides precious few answers. It's no wonder people are confused. Through decades of rigorous research, the extent of what we now know in the field of nutrition and dietetics is mind-blowing. Indeed, humanity's insight into the science of what we eat continues to grow at an exponential rate. This is a beautiful thing. However, this means that to possess a reasonable level of nutrition knowledge requires you to do some work, a problem exacerbated by the need to wade through the ever-increasing amounts of contradictory information. Moreover, the more our knowledge of nutrition science grows, the more we realise the extent of what we don't know, and it's becoming increasingly difficult to distinguish between truth and pseudoscientific pretence.

When this stems from people having a range of ideas born out of different interpretations of the evidence, this should be celebrated. Unfortunately, however, opinions are too often based on emotionally charged conjecture serving to assert one's dietology. So, how does one distinguish the more credible information from the bad? While science does indeed attempt to provide answers, how many of us are motivated to trawl through hundreds of papers on a topic and then listen to several experts' interpretations of the ever-changing data? Biases abound, the scientific process is slow, fallible, prone to corruption and poorly understood. Worse: even if we were to follow the scientific consensus, what's to say that this isn't wrong and itself subject to bias?

With so much misinformation in nutrition and dietetics, it's scarcely surprising that people are confused about what foods are 'good' and which are 'bad'. Policymakers can't decide: back in the 1970s, we were told that eggs are bad for us, next it's ok to eat them again.[20] If it is, indeed, the case that we're being fed unsound information, does this mean that the motivations of policymakers are politically led and bias-laden? Probably. At least to some extent. But this is far from the whole picture. The majority of the sub-quality information in nutrition policy results from inadequate standards in nutrition research. Nutrition science isn't the clear-cut evidence-discerning machine we think it is, and like any science – though arguably more so than most – it's inherently flawed. I'm not merely talking about nutrition research that's fraudulent, negligent or even biased in the fudging-of-numbers or p-hacking* sense – although, sadly, this is also evident. More that it's hugely hyped. Nutritional epidemiology – i.e. examining diet and nutrition in relation to health, disease and longevity – is difficult and incredibly nuanced. The psychologist and writer Stuart Richie in his book *Science Fictions: Exposing Fraud, Bias, Negligence and Hype in Science* – in which he explores the disturbing flaws that undermine the validity of modern science – notes: 'An incredibly complex physiological and mental machinery is involved in the way we process food and decide what to eat.' Richie points out that, because of this complexity, observational data are prone to enormous bias, trial noise and the vagaries of the human memory, and randomised controlled trials can be tripped up by the complexities of their own administration. Indeed, he refers to nutrition science as 'an extraordinarily hyped field' and goes on to point out that the reason for the extreme hype in nutrition is that 'nutritional epidemiology is *hard*' (emphasis in the original).[21] How many times have you been told about the latest nutritional fad?

---

* p-hacking is the manipulation of data in a research study in order to get the p-value – i.e. the reference marker for statistical significance – to below the 'magic' $p < 0.05$.

So-called 'cutting-edge' information in the media sells stories. Hype is a huge problem in nutrition, it feeds beliefs providing undue validity, adding further emotional bias to the narrative of a particular diet camp. Nutrition hype gives rise to 'food trends', a term I frequently come across. I dislike the term. Its implication relates to what's fashionable in relation to nutrition science, food technology or flavours. The idea is that food companies should follow the latest food fads in their product development and marketing with the sole motivation to increase sales.

What's more, anyone can be a 'nutritionist'. Simply sign up for an online nutrition course by entering your credit card details, then all you have to do is a bit of homework, sit a short test and, *voila!*, you're a 'nutritionist'! There are definitely some excellent online courses available, but unless you already have a reasonable understanding of nutrition science, it's not easy to spot the credible from the quackery. I'm a UK 'Registered Nutritionist', which means I hold an accredited title that demonstrates I'm a qualified nutrition professional that meets a rigorously applied standard for scientifically sound evidence-based nutrition,[22] which includes having studied nutrition and dietetics for four years. The other UK title that indicates the practitioner is the real deal is 'dietitian', a term with the added bonus of being regulated by law.*[23] In the US, 'Registered Dietitian Nutritionist' (RDN) is the accreditation that you should look out for as it provides assurance that the practitioner is a trained expert.[24] However, while these terms provide much-needed assurance that the practitioner is a suitably qualified professional, dietitians are, after all, just human, and, unfortunately, it's not uncommon to find some who misuse their professional authority – intentionally or otherwise – succumbing to their biases and making pseudoscientific claims, falling foul of many

---

* I trained as a dietitian and practised as one in the NHS for seven years from 1995 to 2002.

of the same criticisms I've unleashed on the ultracrepidarians. On the flipside, there are a number of unqualified online influencers who create excellent, evidence-based content: a boon for their followers.

And there's another problem. Legislation isn't always on the side of quality advice. In the UK and EU, a legacy law mandates that health-care professionals (HCPs) aren't permitted to make authorised claims when discussing products. Yet an unqualified influencer isn't shackled in the same way. Article 12(c) of the 2006 EU Nutrition and Health Claims Regulation prohibits authorised health claims by HCPs,[25] such as doctors and dietitians, in advertisements. This includes product labels, print adverts, broadcast, in-store promotions, blogs and social media. HCPs can provide advice on dietary guidelines and public health messages, such as to eat five fruit and veg per day, but they aren't allowed to recommend brands or advise on any health claims. This means you're more likely to hear dangerous advice from a celebrity who's being paid than from a credible impartial expert. While this may be UK and EU legislation, social media is global: people watch videos produced by European HCPs all over the world. Fortunately, there's a backlash and many rightly argue that consumers are better protected by health claims made by professionals who are guided by codes of practice rather than by unqualified, unregulated influencers.[26]

What to Do About It

With these contradictions in mind, how can you tell if an influencer is someone to be trusted or otherwise best shunned? To help, I've provided a list of useful watch-outs to help you spot a charlatan. These traits are not definitive, however, and neither do all pseudo-influencers display all – or sometimes any – of the traits. Too often, dangerous advice is all too obscure.

To Help You Spot a Quack Influencer, Ask Yourself ...

- Do they speak in absolutes, saying things like you 'must' eat a certain food, or that you should 'avoid' something?
- Do they claim what they're saying is 'the truth', or end every sentence with the word 'fact'?
- Are they fearmongering their audience into avoiding certain foods?
- Do they use technically sounding terms like 'toxic', 'inflammation' and 'insulin' without explaining what they mean?
- Is their communication style grandiose or evangelical?
- Does their content imply outlandish promises, such as if you heed their advice, you'll live a long, disease-free life?
- Do they respond to criticism with strawman arguments?
- How transparent are they about their conflicts of interest or financial incentives?
- Are they untrusting and conspiratorial in their narrative, implying that others with nefarious motives are out to get you, but if you heed their advice you'll be safe?
- Do they shun conventional science in favour of an ideology?
- Do they contradict themselves with their own logic in different posts?
- Do they use aggressive or admonishing gestures, such as waving their fist in the air or wagging their finger at the camera?
- Are they egged on by an audience who continually adorns them with praise?

Social media platforms continue to be rife with influencers dolling out dangerous misinformation in the guise of good health. When challenged by experts, zealots tend to double down further on their message. Of course, none of us like to have our beliefs challenged: it makes us feel uncomfortable and it's counterintuitive to consider ideas that run contrary to our beliefs. But an ideologue will remain committed to his or her ideas. If you're a supporter of an influencer who dismisses science in favour of their own made-up narrative, and you're failing to critically appraise their advice, then you're following a dietology. But by using the tools to help you recognise nutrition misinformation, and through being empowered with knowledge of diet and nutrition, you'll be able to spot the charlatans and switch to following accounts that provide sensible advice. Encouragingly, there seems to be a growing number of qualified and experienced nutritionists and dietitians on social media who provide evidence-based, practical tips that they relay in an enjoyable and easy-to-understand style.

Of course, I should address the elephant in the room. As Huel co-founder and the nutritionist behind the products, I have a conflict of interest when I present my views on food and nutrition. Moreover, not only do I have a financial interest, I'm emotionally attached, too. As I've personally approved the nutrition of Huel products, even if I were to no longer work for Huel, it would be hard for me to avoid this bias. All I can do is be transparent and strive to keep this front of mind as I communicate my ideas. This way, any reasonable objection to my ideas should challenge the nature of my argument and the evidence I present, rather than focusing on my biases. Like anyone, all we can do is be aware of our biases, try to mitigate them and, most importantly, be transparent about them.

Angelic Enticements and Magic Bullets

Misinformation doesn't just come from bad actors on social media. Disingenuous marketing practices are older and more endemic, with companies using tactics to lure potential consumers into purchasing their products. The effects of marketing have a greater influence on our choices than we realise: food ads are everywhere. Don't underestimate the effect of food fashions either: gaining insights into food trends is itself a growing industry. With ever-improving online algorithms, 'tailored' advertising is only going to increase the effect of marketing on your subconscious choices. Packaging design influences sales of foods, and controversial nutrition claims – permitted under legislation or otherwise – will affect your purchasing habits. The food-marketing sector is huge and advertising spend is often one of the largest expenditures of a company's budget. Ad spend for food in the US reached $5 billion in 2019, a 12 per cent increase from 2012, with six companies responsible for 70 per cent of the industry's advertising.[27] This serves to indicate how much of an influence the marketing of products has on your choices of food.

When it comes to choosing what to eat, we have our own beliefs, biases and preconceived ideas. These dietologically influenced behaviours need not relate merely to whole dietary strategies, but to food items too. Food marketers are aware that customers are susceptible to the pull of healthfulness and manipulate their labels with health and nutrition claims to reel us in. One such cognitive bias is what psychologists call the 'halo effect'. Here the presence of a health claim induces consumers to rate products higher on perceived health attributes, even those not mentioned in the claim.[28] Another is the 'magic-bullet effect' where consumers attribute inappropriate and exaggerated health benefits to a product.[29] Research has also shown that front-of-pack claims are often sufficient to motivate customers into the unjustified belief that a product is healthy.[30] It's almost as if consuming certain foods invites the divine intervention of cure-all.

Are you distrustful of health-related marketing claims on food labels? Many people are. That's not to say that food labels are always solely out to deceive; many companies include factual information and this can be a helpful way of navigating the complexities of super-market shelves. Being aware of your shopping biases is where contemplative nutrition comes in. Will that breakfast cereal *really* protect your heart? Will an 'immune-boosting' yoghurt *really* help you stave off COVID-19? No single food is a medicine: it's your whole diet that's important.

Part II

Contemplating Our Food Choices

Chapter 4

Contemplating Food and Our Physical Health

My mother, Marilynne, washed two capsules down her throat with some bottled water. She had been drinking carrot juice twice daily for a few years, but the inconvenience of juicing several carrots twice a day – not to mention the considerable amount of washing-up – led her to switch to beta-carotene supplements after hearing that they were equally effective. Her skin had taken on a noticeable orange tinge, but she felt it was worth it due to the anti-cancer protective effects of the carrot juice and supplements.

While she was in the kitchen gulping down pills of beta-carotene, I was a teenager swallowing a couple of capsules called 'arginine pyroglutamate plus lysine'. This was the latest muscle-building ergogenic aid I'd read about, this time in a magazine I'd been sent by Allsports, the company who produced the supplement. I'd only been training a few months and I was desperate to turn my skinny adolescent body into a ripped, muscular physique. The arginine pyroglutamate plus lysine capsules were one of several supplements I'd introduced into my daily routine along with a number of other tonics in my quest for 'more size'. My mum was a little concerned about all these supplements that I was taking and that I seemed to be obsessed with bodybuilding. However, she was pleased that at least I was getting plenty of exercise and was making healthy food choices, and she noticed that weight training seemed to be building my confidence.

Shortly after her diagnosis of breast cancer with a poor prognosis in 1980, my mum had the tumour removed followed by a course of radiotherapy. The procedure had been successful and the doctors were optimistic that her cancer was in remission. However, she felt that she needed to take action to prevent the cancer from returning. She'd come across the Bristol Cancer Help Centre and was following their programme, which included a number of strategies that included a vegetarian diet, relaxation techniques, numerous supplements and occasional retreats at the centre. My mother felt that their programme was useful and it afforded her some control over her condition. The centre had advised on regular use of carrot juice but later suggested that switching to more convenient beta-carotene supplements would continue to provide adequate protection.

These kinds of alternative treatments for cancer lacking the robust science to back up their efficacy were rife in the 1980s. Similarly, the bodybuilding world was steeped in numerous fads about the latest muscle-enhancing supplements. Incidentally, while a quick Google search reveals that arginine pyroglutamate plus lysine continues to be available from a few supplement companies, it's scarcely heard of now* and beta-carotene has been largely discredited as a supplement to protect against cancer,†[1] over the last 30 years little has changed. Manipulative marketing is now even more creative: our cyber shelves are stacked high with the latest 'anticancer' and 'get-ripped' tonics. Maybe I shouldn't be too dismissive: if my mother hadn't been so invested in these sort of treatments, and I hadn't blown my hard-earned pocket money on the latest ineffective muscle-growth

---

* At the time of writing, Allsports still sells it.

† Clinical trials in the 1990s involving people given high doses of beta-carotene supplements failed to support the claim that it's protective against cancer. Indeed research revealed an associated increased risk in certain populations, highlighting the questionable validity of any claimed benefit.

supplement, it's unlikely that I would have developed an interest in nutrition and subsequently pursued a successful career in the field. Both Mum's cancer-combatting regimen and my get-big-muscles-fast strategies did, after all, promote intakes of good-quality nutrient-rich foods.

The nutritional quality of what we eat should be the most fundamental consideration of any dietary strategy. For this reason, it's the first pillar of contemplative nutrition. Putting your own health and wellbeing before that of other people allows you to be in the best position to be able to care about their health and wellbeing. Dietetics, nutrition and food science are massive subjects, and the goal of this section, although centred on nutrition for physical health and performance, is not to discuss the nutritional content of what we eat in any great depth, rather to look at how the nutritional value and quality of food plays into our dietary choices. There are numerous other books and websites that cover the nutrition content of food in greater detail. Here, my goal is not merely to share knowledge, but to empower you to have a more mindful perspective when it comes to spotting reliable information amid the oceans of misinformation. Knowing what good-quality food is can actually be simpler than many people would have you believe. When you realise that eating a nutritionally sound diet really needn't be that hard, you'll be able to focus on the other pillars of contemplative nutrition, safe in the knowledge that you're eating a good diet.

With this in mind, here I will focus on just a few of the important topics in relation to nutrition and physical health, including weight control, some of the controversies about the types of fat we should be eating and why fibre is a too often overlooked nutrient. This chapter will also cover food choice in relation to exercise and physical activity. But first, let's look at what's meant by a poor diet and remind ourselves that we're fortunate to be able to have the nutritional value of food as a consideration when we're choosing what to eat.

Nutrition: Problem vs Privilege

When hearing the term 'malnutrition', images of starving children in Africa likely come to mind. Correctly, the term means 'poor nutrition' and refers to any diet that fails to provide the right amount of nutrients. 'Malnutrition', therefore, refers to both undernutrition – including wasting, stunting and vitamin and mineral deficiencies – and overnutrition, i.e. having too much of certain nutrients, like sugar, salt or calories. Obesity, therefore, is a type of malnutrition.

Americans spend about 6 per cent of their income on food consumed at home and Brits spend about 8 per cent. Compare this to China and India where people spend around a quarter and a third respectively. At the other extreme, Nigerians fork out a whopping 56 per cent of their income to feed their families.*² Although this is primarily due to the fact that, as income rises, food spend falls proportionally, it's also linked to rapidly declining food prices in most Western nations. Consequently, those of us living in affluent nations should acknowledge how privileged we are to have the nutritional content of our food as a consideration when deciding what we want to eat. Those who are less wealthy don't have this luxury. Not only do we benefit from scientific advances that endow us with the know-how to improve our health, wellbeing and performance, we live in a time of plenty where adequate food is readily available for most of humankind. As I touched on in Chapter 1, advances in science, technology, agriculture and logistics are responsible for this. Of course, there are still far too many people with limited access to adequate sustenance, but this is generally a consequence of war, politics or social factors. No human in the third decade of the 21st century should be undernourished – we have the capability of feeding everyone.

* Based on GDP per capita.

Of the 11 million deaths per year directly linked to poor diets, non-communicable diseases (NCDs) like diabetes, cardiovascular disease (CVD) and obesity are just some examples.[3] 'Suboptimal nutrition' – the term used by scientists that refers to dietary risk for diseases linked to either hunger or glut – affects 255 million more lives through disability as a result of chronic disease linked to diet.[4] Nearly one in four people across the world are classed as 'overweight' or 'obese',[5] and, according to the Centers for Disease Control and Prevention (CDC), nearly 42 per cent of Americans were overweight in March 2020, more than 9 per cent of whom were severely obese,[6] and the estimated medical cost of adult obesity in the US reached $173 billion in 2019.[7] Looking in the other direction, 800 million people globally are estimated to be undernourished.[8] The charity Health Poverty Action claims that 'Globally, two billion people suffer from malnutrition in one form or another'.[9] From a global perspective, eating too much is a bigger problem than having too little to eat: a position that's flipped as recently as during the last 30 years.[10] Consequently, we live in a time when an increasing number of people pay attention to the nutritional value of what they put in their mouths, at least to some degree. Even when they regularly consume junk food, most people do so knowing that it might not be great for their health in the long term. They are, therefore, considering its nutritional content – or lack thereof – even though it bears little influence on their motivation to eat it.

Assuming someone has the ability to access nutritious food, their meal choice will relate to their own priorities. The majority of people are aware that what they should be eating is beneficial for their health – or, at least, unlikely to harm them – and that nutritious foods should be present in their diet at least sometimes. Rather, at a minimum, they *aim* to consume what they *perceive* to be good for them. Some of us choose food based on our goals relating to gaining muscle, increasing strength or improving physical endurance in order to optimise our performance for a particular sport or pastime. Those who are concerned about their weight make their food choices with the objective of reducing the amount of fat they hold. Others have an allergy, intolerance or autoimmune disease linked to a particular food or food group, and, for these, it's imperative that they avoid these foods.

We're not always provided with reliable information, even when it comes from seemingly trusted sources. As a consequence, what we choose to eat might not always be as nutritionally beneficial as we think. However, we should remember that most of us are living in the most affluent time in history and are enjoying greater access to a wider range of foods. The food system, as we're seeing, is far from ideal and possibly the biggest health concern concerning nutrition in current times is excess body weight.

A Weighty Issue

Popular belief tells us that when the amount of energy (measured in calories) we consume is greater than the amount we use, we'll gain weight by storing body fat. Of course, according to the laws of physics this is an indisputable fact. However, in the real world, the claim remains true only to a point. When it comes to gaining weight, the physiological processes involved are extremely complex: multiple processes regulate both the energy we consume and the amount our bodies burn. When we add psychological influences to the mix, such as our relationship with food and how we deal with cravings, socio-economic factors and the effects of genes, among other influences, things start to get really murky.

One of the biggest problems in the modern Western world is society's continued insistence on focusing on body weight and judging people on how they look. In the age of social media, the persisting negative stigma surrounding being overweight is staggering and only seems to be getting more pronounced. This anti-fat mindset is confounded by the incorrect assumption that people who carry excess fat are unhealthy and lazy, a mentality that exacerbates a vicious cycle where people who struggle to control their weight are negatively pre-judged, further hindering their attempts to rectify their situation. However, as we've seen, multiple factors influence what foods we eat, and numerous mechanisms – biological, social and cultural – affect how our bodies utilise energy, including our ability and motivation to perform physical activity. When considering this, we should ask ourselves: *is how much excess body fat we hold purely a matter of choice?*

BMI – A Bad Measurement for Individuals

Body mass index (BMI) is a crude measure of body weight for height. It was originally devised by Adolphe Quetelet, a Belgian statistician, in the mid-1800s.[11] Using data collected from 19th-century white European males, Quetelet aimed to establish what an 'ideal' weight was. Originally known as the 'Quetelet Index', it was renamed 'body mass index' in 1972 by American physiologist Ancel Keys, from which point its use became widespread.[12] A BMI value of below 18.5 is classed as 'underweight', 18.5 to 24.9 'normal weight', 25 to 29.9 'overweight' and 30 and above is 'obese'.

BMI is now the dominant metric used by medical practitioners, health researchers, life insurance companies and the food industry. It's used to predict health outcomes and disease risk at the population level. However, when used to predict the health of individuals, as it often is, BMI is flawed. How does something that measures the average weight of white middle-class European males 170 years ago bear any accuracy as a predictor for all ethnicities, genders and socio-economic groups across the world today? And that's not to mention its limitations in respect of athletes and those who carry greater amounts of muscle tissue. BMI may have some use in epidemiology when employed as a statistical measurement for groups and alongside other anthropometric measurements in relation to disease outcome. But an overreliance on BMI as a marker for what's considered 'healthy' means that 'normal weight' bodies are seen as the benchmark for health. This gives rise to selection bias when considering eligibility criteria for clinical trials that produce results that go on to influence standards of care.[13]

Despite there being numerous biochemical, performance and other anthropometric measurements used to assess people's health, BMI is too frequently the principal tool used to decide whether someone is eligible for a treatment. For example, doctors can only prescribe the satiety-promoting weight-loss drug semaglutide to patients with a BMI of 35 or more.*[14] Might it not be more beneficial to proactively treat someone with the drug earlier so as to prevent them from gaining too much weight in the first place, at least in some cases? Similarly, a patient with a BMI that's too high might be excluded from critical knee surgery that would help their mobility, enabling them to be more physically active.

The term 'obesity' itself fails to convey the complexity of the condition, which manifests in dysregulated dietary intake, adipose tissue and metabolism. Because the nature of obesity centres on mechanisms that regulate dietary intake, rather than on BMI – which is how people generally define it – by altering our focus, we can apply more effective treatments. However, it was somewhat controversial when, in 2013, the American Medical Association classified obesity as a disease.[15] This action intended to send a clear warning about the manifold health risks associated with being overweight, and hoped to

---

* Or 32.5 or more to people of Asian, Chinese, Middle Eastern, Black African or African-Caribbean descent.

change the way the medical industry views and pays for treatment, aspiring to destigmatise the condition. However, while obesity is a risk factor for many diseases, not everyone with obesity suffers from ill health or displays adverse metabolic markers (such as high cholesterol, raised blood sugar or high blood pressure) known as metabolically healthy obesity (MHO).[16] Nevertheless, efforts to educate the public about obesity often ignore why the disease has become so common, instead treating individual behaviour change as a cause and solution. Treatment should refocus on getting sick people well, not making 'fat' people thin. In fact, the authors of a 2023 paper have gone as far as to propose obesity is reclassified as 'adiposity-based chronic disease', because, they say, it's not about treating an elevated BMI, rather a disease influenced by environmental, genetic, physiological, behavioural and developmental factors.[17] Viewing obesity this way would allow more effective treatment, especially as bariatric surgery and drugs like semaglutide typically lead to significant weight loss and improvements in metabolic health.

Pulling in the Wrong Direction

Along with the shortcomings of BMI, let's turn our attention to body weight. To a large extent, we have little control over how much we weigh. To a surprisingly significant degree, body weight is determined by our genetic make-up. Research involving twin and adoption studies has led to suggestions that as much as 70 per cent of our body weight is down to our DNA.[18] This heritability of body weight encompasses numerous factors such as metabolic rate, appetite and drive to eat, mindset, motivation and our ability to exercise, as well as less obvious factors like our behaviour towards food and physical activity. The huge influence of genetics on how much body fat we hold flies in the face of what many people believe, and highlights the enormity and complexity of the link between genetics and the environment in determining how heavy we are. The ability to lose weight is more than simply a matter of willpower and goes some way to explain why some of us find it relatively easy to stay lean and others struggle to shift a pound off the scales. So, if 70 per cent of our weight is the legacy of our genes, what about the remaining 30 per cent? This is where we have the ability to exhibit control, meaning that losing weight is possible for almost everyone, albeit through varying degrees of effort.

In his book *Why We Eat (Too Much): The New Science of Appetite*, Andrew Jenkinson describes observations which, at first glance, seem contradictory.[19] He notes that between 1980 and 2000, Americans were, on average, consuming an additional 500 calories per day.[20] Because, broadly speaking, consuming an additional 7,000 calories equates to roughly 1 kg (2 lbs) of body weight, he calculated that over one year – i.e. 500 kcal × 365 days – the weight gain of the average American man should amount to 26 kg (57 lbs). Puzzlingly though, available data suggest that during this period the actual weight gain for the average American consuming an additional 500 kcal per day was only 0.5 kg (1 lb) in a year. The maths doesn't add up. How is this possible? What magic formula made those extra calories disappear?

According to Jenkinson, this seemingly fictional phenomenon has little to do with magic and more to do with our biology. In the book, he argues that our bodies cannot hoard an infinite amount of energy and so when food intake increases our metabolic rates go up so as to burn more energy. The general idea behind this theory is that human body weight is tightly controlled by a system of checks and balances that keeps your metabolism at a steady-state level, or 'set-point'. This 'weight set-point theory' originated from a series of animal studies conducted during World War II by a couple of undeterred scientists called Hetherington and Ranson. Their experiments showed that mechanical injury to an area of the brain, known as the ventromedial hypothalamus, leads to obesity in rats whereas similar lesions in the neighbouring lateral hypothalamus causes rats to lose their appetite.[21] In both cases, the lesions prevented the animals from bouncing back to their original body weight, even when forced to eat more or less food. This research gained momentum in the late seventies when another group of scientists showed that feeding rats with an energy-dense diet results in overeating and a disproportionate weight gain.[22] However, when again exposed to a mixed diet, the rats spontaneously returned to their original weight – as though genetic pressures were pulling them back towards their prior set-point.

Unlike rodents, studies in humans have provided less clear-cut evidence. This is not for lack of trying. Studies linking metabolic set-point to body weight in humans date back to the 1940s with the Minnesota Starvation Experiment when Ancel Keys (who renamed the Quetelet Index to BMI) and colleagues explored the effects of starvation on people's metabolism. Their results showed that 24 weeks of semi-starvation (i.e. 50 per cent reduced energy intake) caused participants to lose 66 per cent of their initial fat mass but *ad libitum* re-feeding resulted in a regain of fat mass reaching 145 per cent of the pre-starvation values.[23] To the delight of many scientists working in this area, the Minnesota study was the first to demonstrate that when you change someone's diet, their metabolism will adapt as a result. Since its publication in 1950, the study has been heralded as key evidence in favour of the set-point theory, partly due to the fact that long-lasting starvation studies in humans are hard to come by for obvious ethical and practical reasons. Using slightly less gruelling approaches, research in the 1980s and 1990s seemed to support Keys' findings. Observations in humans showed that when the system is perturbed – for example, by a period of dieting or overfeeding – people lose or gain weight, but when they stop sticking to their diets, they start to pile on the pounds or they lose accumulated fat, returning to their original weight.[24] Echoing Keys' observations, some even showed that as your body slims, your metabolism plummets to resist alterations to your diet: it's as though your body shuts down energy expenditure to activate a metabolic switch, which curbs further weight loss. As far as the set-point theory goes, this biological feedback ensures that your weight never falls too far from a genetically determined weight range. A series of biological mechanisms influence both our metabolism and sensations of hunger in an attempt to keep our body fat at a particular level and to compensate for any weight change.[25] The model revolves around the idea that your weight may go up or down temporarily but will ultimately return to its normal set range – think of it as the weight at which our body is most 'comfortable', or, as described by Jenkinson, 'like the thermostat that controls the temperature in our house. It will reach, and then maintain, the level that it is set at'.[26]

Despite its obvious appeal, there are several aspects of the set-point model that are rather problematic. The idea accommodates many biological aspects of energy balance – for instance, the energy-efficient role of leptin in placating hunger – but fails to embrace many environmental and social aspects of eating behaviours. In a world of abundance, it seems likely that even in the presence of an active biological control of body weight, any homeostatic signal would be hijacked and sometimes overshadowed by external cues. Our drive to eat is not only regulated by visceral, biological signals, but also by powerful, attention-grabbing environmental clues. The vast availability of food choices in rich-world countries showers our brains with an overwhelming number of mouth-watering cues, inducing our bodies to store increasingly larger amounts of energy. In this context, the set-point theory does not address the possible gene–environment interactions: why does your waistline expand during a pandemic lockdown or after you get married? Or why does car ownership make it harder for you to zip up your jeans?[27] Today, gene–environment interactions are beginning to emerge as powerful drivers of obesogenic behaviours, raising questions around the usefulness of 'gene first' approaches to study weight control.[28] In addition, a recurring theme in human studies is that most results come from observations in free-living subjects where external factors may camouflage any evidence of active biological control.[29] In these studies, socio-economic drivers are also likely candidates contributing to this tangled tale: studying overweight subjects in affluent societies is unlikely to address biological control (self-control is more likely to be at play). In short, observational studies in humans do not provide consistent evidence for a biological control of body weight, which, if it exists, may be overridden by environmental and cultural influences on behaviour. This implies that searching for the genetic determinants of obesity in a world of abundance is self-defeating: for all the research, living near a McDonald's may easily skew any genetic variability in weight gain.[30] Instead, an alternative model better describes what's going on. The 'settling point' concept suggests our weight is influenced by numerous other factors rather than merely our genes. How we navigate our food choices together with our biological traits and energy balance affects how weight shifts over time.[31]

A Big Fat Controversy

The extent to which being obese is detrimental to health has become something of a controversy. Although body weight is influenced by numerous factors, the traditional view that suggests being overweight is suboptimal for our health is not shared by all health professionals. One example of an alternative perspective comes from Dr Joshua Wolrich who, in his 2021 book *Food Isn't Medicine: Challenge Nutrib*llocks & Escape the Diet Trap*, discusses the multiple influences on energy balance. Wolrich – a doctor in the UK National Health Service (NHS) and a social media influencer with a considerable following – claims that society has an unduly maligned view of

body weight and obesity and this negative stigmatisation has contributed to unnecessary suffering. According to Wolrich, not only is it
okay to be overweight, but it can be perfectly healthy.[32] Humans are
hard-wired to resist weight loss, and because of this, he claims, it's
perfectly acceptable to be overweight, stating that 'weight loss has
never simply been a matter of willpower'. He continues: 'Treating it as
such turns our body weight into a personal responsibility, which in
turn leads to overt discrimination.' He references studies that indicate
healthier BMIs are, in fact, those that would classify an individual as
'overweight'.[33] He makes a very valid point: the 'obesity paradox' indicates that some NCDs commonly associated with excess body weight
have a lower risk of poor outcome at BMIs above 'normal', and
research has shown that BMIs of up to 28 or even 31 (i.e. in the 'overweight' and 'obese' categories) may even be associated with lower
mortality in people with type 2 diabetes and CVD.[34]

Wolrich's argument is well founded. Society does stigmatise those
who are overweight who we too readily consider to be 'unhealthy',
despite the data being far from clear. He correctly points out that the
biology of body weight is highly nuanced and involves multiple influences: you can, for instance, be 'overweight' and physically fit. Indeed,
the prevalence of MHO could be as high as 60 per cent of 'obese'
adults in America, depending on how it's classified, which calls into
question whether obesity should even be considered a disease.[35]
Should people be diagnosed with a disease or be viewed as 'abnormal' or 'unwell' when they're not sick? What concerns me, however,
is that Wolrich's perspective does little to acknowledge the multitude
of research that shows an association between higher BMI and
disease risk in populations. For example, numerous studies have
linked obesity to CVD risk,[36] several cancers[37] and diabetes,[38] and
there's demonstrated associations between increased risk of ill health
and mortality.[39] And, as well as increased risk of CVD, cancer and
diabetes, being overweight contributes to loss of mobility with age[40]

and is associated with lower physical activity levels.[41] Of course, multiple factors come into play and I share Wolrich's criticisms of BMI,[42] but I'm deeply cautious of his bias: he includes an account of his own experiences of being overweight and having an unhealthy relationship with food which, he claims, has been psychologically damaging.[43]

Obesity is a recent phenomenon as demonstrated by the absence of evidence that our hunter-gatherer ancestors were significantly overweight, and there's been little change to the human genome since the agricultural revolution.[44] Neither has obesity been observed in members of modern hunter-gatherer tribes[45] nor those residing in traditional agrarian societies,[46] at least to any notable degree. Moreover, obesity has increased in recent decades with the worldwide prevalence more than tripling between 1975 and 2022.[47] It's reasonable to conclude, therefore, that being obese is not the human default state and that factors relating to our modern lifestyle with an abundance of food has contributed to the epidemic. While we should accept that you can be overweight (as defined by BMI) and in good health, this is contingent on regular physical activity and making sensible food choices, among other influences. Moreover, we should be mindful of how the evidence is interpreted: a 2019 systematic review and meta-analysis, for example, indicated that the obesity paradox in MHO isn't always apparent, noting that for every unit increase of BMI above 20 there was an increased risk of CVD.[48]

Using BMI as a marker of health is flawed and Wolrich is right to challenge the mainstream medical presumptions made by doctors. A BMI of over 25 certainly doesn't mean you're 'unhealthy' and we should be encouraging people to focus less on their weight and more on their relationship with food. Society shouldn't be prejudicing any weight, body size or shape: being obese is the result of a complex interplay of numerous genetic, physiological, socio-economic and psychological factors over which we have little control. However, any dismissal of the health concerns associated with being overweight or obese should be challenged, otherwise, I fear, we're edging dangerously close to the postmodern 'fat studies' position where being 'fat' is solely an identity, and where even health-related claims in relation to body weight are seen as mere social constructs.[49] I agree with Wolrich that when we discuss BMI and health risk we must be cautious of our biases and watch out for cherry-picked research,[50] and that we should be educating the public and health professionals that obesity is rarely a choice. But I'm concerned that Wolrich's perspective in *Food Isn't Medicine* – one that's shared by other professionals – fails to sufficiently acknowledge the dire health risks associated with living with obesity. We should be encouraging weight control through positive interventions and highlight the benefits of exercise and healthy food choices that go beyond fat loss, while, at the same time, positively supporting people in their weight loss journey. We should certainly not be admonishing anyone for 'not trying hard enough'.

Convenient Answers to Complex Issues

You may have noticed that a number of people like to work out precisely how many calories or grams of protein they require in a day and how the foods they eat contribute to their daily intake of energy or macronutrients. We humans like convenient explanations for complex things and, while some people find this useful, for others, having to calculate our intake of several nutrients can be a load of unnecessary hassle. Nutrition science rarely provides such simplistic descriptions, and numerous examples can be found in food and dietary policy. To illustrate the complexity of such issues, I'm going to focus on three topical areas, all relating to dietary fats: trans fats, omega-3s and saturates. I choose these three examples because they illustrate perfectly that giving dietary advice and devising food policy off the back of research is far from straightforward. If we don't know what an optimal diet looks like – despite knowing what a poor diet is – then, while being mindful to avoid unhealthy foods, we should emphasise other perspectives. This is an underlying theme of contemplative nutrition.

Confounding by Indication

Nutrition scientists have a tough job. They are continually faced with multiple obstacles in their attempts to reveal disease-causing nutrient culprits and then having to translate these into dietary advice. A useful illustration of this arduous task of trying to extrapolate a nutrient factor from epidemiological data and its associated NCD risk relates to trans fats in the 1980s and 1990s. Trans fats are types of fat that have a different chemical configuration to the more common cis isomeric fats that we consume. They are produced when oils undergo a process called hydrogenation, which used to be prevalent in the manufacture of certain foods, particularly margarine and certain confectionery items. Trans fat consumption has been associated with increased risk of a number of NCDs, in particular CVD, but also diabetes, cancer, certain allergies and obesity.[51] Thankfully, through considerable consumer pressure and legislation, trans fats are rarely present in more than trace amounts in our foods. This is one of those rare moments in nutrition where all camps are in agreement: trans fats are unhealthy and should be avoided. A plethora of evidence has revealed their danger, but this hasn't always been the case. A landmark 1990 paper published in the *New England Journal of Medicine*, by Dutch scientists Ronald Mensink and Martijn Katan, revealed that intakes of these fats worsened the blood lipid profile by increasing harmful LDL cholesterol and lowering protective HDL cholesterol levels.[52] Following the Mensink and Katan paper, a 1993 *Lancet* report by Walter Willett and colleagues at the Harvard School of Public Health in Boston demonstrated a positive association between trans fatty acid intake and heart disease risk.[53]

Although the health dangers associated with trans fats now seem glaringly obvious, this hasn't always been the case, and following a number of reports on the dangers of trans fats, in 1997 one epidemiologist, Samuel Shapiro, warned of a possible bias, which he referred to as 'confounding by indication'.[54] As, prior to the 1990s, public health policy had been promoting the health benefits of margarine over butter, many individuals with prior coronary heart disease or who had been identified at risk of CVD may have been more likely to have followed official health policy and to have opted for margarine, and therefore had a higher intake of trans fats. If this were the case, then the positive association of trans-fat intake and CVD risk may have been exaggerated because the people in this high-risk group had altered their behaviour to one which, at the time, they believed to be a positive behavioural change. Shapiro pointed out that this did not necessarily mean that trans fats contributed to CVD. Of course, now we have mountains of evidence demonstrating that it is, indeed, the case that trans fats have dangers associated with their consumption. So, why do I bother to mention this now seemingly moot issue? It serves as an example of excellent scientific practice; something all too rare in nutrition policy. One researcher challenges a narrative that is rapidly gaining momentum, highlighting a potential watch-out in nutrition science even when evidence points to something. This is a bias hurdle that needs to be navigated when providing dietary advice relating to other nutrients, as we shall see.

The Omega Ratio

Over the past few decades, the nutrition policy of several nations has recommended that people include oily fish – for example, salmon, sardines, pilchards, mackerel and trout – two or three times in their weekly diets. These foods provide omega-3 fatty acids, a good intake of which may protect against some NCDs. Sometimes these recommendations go further, noting that, if you don't consume oily fish, you should consume algae or fish-oil supplements. As well as this, some experts claim that it's not just consuming omega-3s that's important, but that we also need to consider the ratio of omega-3 to omega-6 fatty acids, meaning that we possibly should also be moderating our intake of the latter. Let's explore the validity of these claims, as most of us just want to enjoy nutritious food without worrying about the numerical ratio of one nutrient to another.

One of the categories of dietary fat are polyunsaturated fatty acids (PUFAs), and the two main types of these that we consume in our diets are omega-3s and omega-6s (the terms refer to their structure). As humans, we must include two essential fatty acids (EFAs) regularly in our diets: linoleic acid (LNA), which is an omega-6, and alpha-linolenic acid (ALA), an omega-3. In addition, although they can be synthesised in the body, two other omega-3s can be useful to consume as they reduce the requirement for ALA: eicosapentaenoic acid (EPA) and docosahexaenoic acid (DHA). EPA and DHA can be obtained from oily fish, marine algae or supplements. If you don't eat oily fish, you'd be wise to ensure that you acquire a good intake of omega-3s from other sources – like flaxseed, chia and hemp seeds, walnuts or some vegetable oils. Omega-3s are also present in smaller quantities in a wide range of plant and animal foods. The levels of EPA and DHA in the blood are linked to both how much we have in our diets and the efficiency of certain biological processes involving their conversion from ALA.[55] There is, however, relatively inefficient conversion, and, for this reason, if your diet doesn't contain much EPA and DHA, you'll need a considerably higher intake of ALA. A good level of omega-3s in the blood – especially EPA and DHA – is important due to protective effects in relation to CVD and other conditions where inflammatory processes are involved.[56] In addition, low levels of DHA have been linked to negative neurological effects.[57] Similarly, there are two semi-essential omega-6 fatty acids: arachidonic acid (ARA) and gamma-linolenic acid (GLA), and a good dietary intake of these reduces the requirement for LNA.

The requirement for omega-3s has been shown to go up if a diet is higher in saturated fat, trans fat or alcohol, or if there's a high level of insulin in the blood.[58] These influence the function of the enzyme that converts ALA to EPA and DHA. Also, so the theory goes, diets high in omega-6s have been shown to lower the conversion of ALA to EPA and DHA because ALA competes with LNA, which is relatively abundant in many diets, as the same enzymes are involved in converting LNA to ARA.[59] This implies that a diet rich in omega-6s increases the demand for ALA in the absence of EPA and DHA and therefore the ratio of omega-3:omega-6 might also be important.

One of the reasons that our bodies require fatty acids – both omega-3s and omega-6s – is because they make up the structure of cell membranes. The strength of the membrane depends on a number of factors, one of which is the degree of presence of omega-3s. The amount of omega-3s that's taken up depends on what's available in the blood, and, it's been suggested, if there are too many omega-6s, the membrane becomes weaker and less resilient.[60] The greater the amount of omega-3s available relative to the amount of omega-6s, the more strength-building omega-3s will be incorporated into the cell membrane. The inflammatory response is also relevant. Eicosanoids are signalling molecules involved in the immune response and are linked to the degree of inflammation. Omega-6 fats produce more pro-inflammatory eicosanoids and omega-3s more anti-inflammatory ones.[61] Again, omega-3s and -6s compete with one another: they share some of the enzymes involved in eicosanoids production. This means that an imbalance in the ratio of omega-3 to omega-6 may also lead to an imbalance in the inflammatory signalling molecules.

In order for the ideal quantity of each type of PUFA to be incorporated into cell membranes, it's been suggested that the balance of omega-3:omega-6 should be no more than 1:4 (i.e. up to four times more omega-6 than omega-3).[62] It's claimed that our hunter-gatherer ancestors, through acquiring their nutrition from a range of foods, would have had these sorts of intakes, and this is also the case for people living in remote areas today. Intakes of many modern diets, however, have led to large amounts of omega-3s being omitted and omega-6s are more prevalent, leading to typical ratios of 1:15 to 1:20 or even higher.[63]

However, the advice that we should stick to a set omega ratio is way too simple. Yet again, we have an example of humans wanting convenient explanations for something complex. The narrative provides no differentiation between particular omega-3 and omega-6 fatty acids, and a higher level of LNA doesn't mean more ARA (the prerequisite of pro-inflammatory eicosanoids) is produced, and even if they did, ARA also forms anti-inflammatory molecules.[64] Moreover, the presence of ALA provides no indication as to whether more EPA will be produced or more DHA. It's questionable enough that people rely so much on counting calories and things like how many grams of protein or carbs are in their meal, without adding further complexity to ensure that for every four grams of omega-6 they consume they have to counterbalance with at least a gram of omega-3. Worrying about the ratio of one set of nutrients to another is extremely reductionistic and unviable. The evolution of human dietary requirements simply didn't require our bodies to have anywhere near this level of detail. While omega-3s play a key role in modulating the inflammatory response to a clinically relevant level, it doesn't mean that omega-6s have the opposite effect. Indeed, many experts feel we should be encouraging whole sources of both types of PUFA.[65]

So what's the answer? Fundamentally, it's important to consume good sources of omega-3s. In traditional diets, this was easier. For example, in pre-intensive farming, the amounts of omega-3s in many plant foods were likely slightly higher and, while on an individual food basis, this was of low significance, over the course of a week, this led to a reasonable contribution to omega-3 status. I'm hoping that this might be something that regenerative agriculture and new technologies can address.

The nutritional quality of the fish we serve up has also changed. Take tuna, for example: the most commonly consumed type of tuna is yellowfin – what you get when you open a can or buy from your local supermarket fish counter. Canned tuna is no longer included in official oily fish recommendations, because, some claim, the canning process removes much of the omega-3s. The canning process is only part of the reason, however: canned tuna may have an unimpressive 0.04 g omega-3s per 100 g, but tuna steak isn't much more exciting at 0.11 g. Compare this to salmon, which is around 3 g. But this hasn't always been the case. Prior to commercial fish farming, free-swimming fish used to feed off plankton which, itself, fed on omega-3-rich algae, meaning that the fish were omega-3 rich. Interestingly, bluefin tuna – an expensive and rare delicacy – is considerably higher in omega-3s: 2.8 g per 100 g.[66]

While the omega-3: omega-6 ratio excites scientists, when it comes to providing nutritional advice, things are way more nuanced. Simply make sure that you're including plenty of omega-3-rich foods rather than worrying about omega-6s. Be careful of confounding by indication.

A Saturated Debate

Recommendations that we should be reducing our intake of saturated fat and, instead, consuming more unsaturated fat has become the cornerstone of nutritional advice. The guidelines of several governments have authorised marketing messages like 'Low Saturated Fat' or 'Low in Saturates', and the saturated fat content of food items is included in the nutrition tables of food labels as one of the 'Big 8'.* Consequently, saturated fat has become so demonised that, for many of us, alarm bells go off when we hear the term. However, as you might expect, the reality is far from straightforward. In fact, there's a divide among experts as to the extent that saturated fat is harmful to health. In February 2020, several nutrition scientists compiled a consensus statement and wrote an open letter to the US Departments of Agriculture and Health and Human Services claiming: 'There is no strong scientific evidence that the current population-wide upper limits on commonly consumed saturated fats in the US will prevent CVD or reduce mortality. A continued limit on these fats is therefore not justified'.[67]

In fact, this controversy began way back in the late 1960s when a British nutrition scientist, John Yudkin, demonstrated a link between sugar intake and heart disease, which he outlined in his now famous book *Pure, White and Deadly: How Sugar Is Killing Us and What We Can Do to Stop It*.[68] In reaction to this potentially damning hypothesis, the sugar industry, it's claimed, donated large sums of money to leading Harvard scientists to help them conduct research and exonerate sugar in favour of shifting the blame to another culprit, i.e. fat. This research trickled down into the mainstream and the 'dangers' of fat, saturated fat and cholesterol became public knowledge and the

---

* 'Big 8' refers to the eight main values listed on food labels: calories, fat, carbohydrate, protein, fibre, sugar, salt and saturates.

rest is history. This wasn't revealed until 2016 when a paper titled 'Sugar Industry and Coronary Heart Disease Research: A Historical Analysis of Internal Industry Documents' was published in *JAMA Internal Medicine*.[69]

Fat-blaming was further exacerbated following the Seven Countries Study led by Ancel Keys (yes, him again). Keys pointed the finger directly at saturated fat and dietary cholesterol,[70] and was highly critical of Yudkin. In the study, Keys' findings illustrated a high degree of correlation between the amount of dietary fat and heart disease in seven different nations. However, what was not included in the original paper, but was later revealed, was that the study looked at, not seven, but 22 countries – the data from the other 15 nations was omitted. Keys, it's claimed, only reported correlations that supported his hypothesis, with, for instance, France and Germany – two large nations that consumed a lot of saturated fat but had lower rates of heart disease – being excluded from the write-up.

There has been considerable research over the past few decades, much of which corroborates Keys' hypothesis. Indeed, multiple well-designed studies have demonstrated that replacing saturated fat with polyunsaturates both improves health markers and reduces NCD risk.[71] Yet not all have arrived at the same conclusion, including a 2017 meta-analysis by Steven Hamley, which demonstrated that replacing saturated fats with polyunsaturated fats had no effect on heart disease.[72] Could it be that some trials – not just the Seven Countries Study – were bias-driven and erroneously designed, and their outcomes over-hyped when reported in the media? Stuart Richie, in *Science Fictions*, notes that 'the only thing that should be changing is the value of interest – in this case, whether saturated or unsaturated fats were eaten'.[73] Changes to anything other than this, like other diet-related disparities, would likely have affected the results. Hamley points out that previous meta-analyses – ones that several governments had based their nutritional advice on – simply hadn't noticed the study design problems.[74]

So we have evidence concerning saturated fats and their effect on health pointing in both directions. This is extremely confusing. However, there's one useful learning to come out of this controversy: it demonstrates the highly nuanced nature of nutrition science, and we're reminded that simple recommendations are rarely feasible. The debate regarding saturated fat intake and CVD risk is an example of the complexity of nutrition research and serves to highlight the difficulties when dietary policy is made off the back of it. The legacy paradigm that 'saturated fats poison the body' may not be as rock solid as some think.

Relevant to this controversy is the so-called 'hunter-gatherer paradox'. Meat-rich diets have been linked to increased CVD risk and the saturated fat content of such diets is often the accused offender. But the reality is far from this simple. Some modern hunter-gatherer populations have been shown to consume large amounts of meat yet have low levels of atherosclerosis, hence the so-called 'paradox'.[75] These communities are also incredibly active, highly cooperative and enjoy plenty of fresh foods, demonstrating that forming a causal link between poor metabolic health and saturated fat intake is epiphenomenal. More likely, poor heart health results from a combination of the multiple impacts of modern living, such as sedentary lifestyle, prolonged stress, reduced sleep and poor food choice. Although humans don't need to eat anywhere near the amounts of meat that many of us do, this doesn't mean saturated fat is a problem. The human body is complex, which makes drawing conclusions somewhat troublesome. I'd argue, in fact, that the hunter-gatherer paradox isn't even a 'paradox', only appearing as one through being linked to biased data from the outset. Indeed, other research has called the link between meat consumption and CVD into question.[76]

So, where do we stand on saturated fat, especially in the absence of a firm conclusion? I certainly don't think that freely consuming copious amounts of saturated-fat-rich foods – particularly at the expense of unsaturates – is risk-free. However, neither do I think that one of the key focus areas on dietary policy in the 2020s should be on saturated fat, especially considering the complex aetiology of CVD. Picking up on a single nutritive correlation and converting it into official policy seems far too simplistic in this case. I certainly don't think there should be an absolute recommendation, like, for instance, limiting intake of saturated fats to 11 per cent of total energy. Despite this, both the UK and US recently updated their recommendations on targeting saturated fat intake.[77] The saturated fat controversy is another example of humans liking straightforward answers to complex issues. Linking one nutrient and a disease outcome is overly simplistic.

What can we learn from this? Exercise caution when attempting to draw a conclusion from a single potential causal agent. The economist and social theorist Thomas Sowell wisely observed, 'One of the first things taught in introductory statistics textbooks is that correlation is not causation. It is also one of the first things forgotten.'[78]

Fabulous Fibre

As we've seen, people have different opinions when it comes to which types and how much fat we should be including. When it comes to protein, we seem to be constantly reminded to have plenty of high-quality sources to grow and repair our muscles, maintain a strong immune system and manufacture crucial hormones and enzymes. And if there's one nutrient that will surely divide the room, it's carbs: one side tells us that carbs should be the mainstay of each and every meal and the other warns us to avoid them like the plague.

Fibre, however, gets way too little attention. Other than being told by our parents to eat our greens, favouring fibre is just not sexy.

Fibre is a catch-all term for all of the carbohydrate compounds found in food that the enzymes of the human digestive system can't break down. There are hundreds of different types of fibre, but we categorise them into two main groups: soluble and insoluble. Soluble fibres dissolve in water and examples include plant pectin and gums, whereas insoluble fibres, such as plant cellulose and hemicellulose, don't dissolve in water. The benefits of a good intake of fibre are numerous: it helps to normalise bowel movements, maintain a healthy cholesterol profile, normalise blood-sugar levels and, because fibre-rich foods promote satiety, maintain a healthy weight. Some types of fibre also help to support a healthy gut microbiome – those friendly bacteria and other microorganisms that reside in our digestive systems – which, as you'll see in the next chapter, can have important implications for our mental health.

The problem, however, is that most of us aren't consuming enough and some public health officials even claim that we're in a fibre crisis. According to the British Nutrition Foundation (BNF), only 9 per cent of UK adults are getting the recommended 30 grams per day, with most adults failing to meet the amount that a five-year-old needs.[79] In the US, only 5 per cent of men and 9 per cent of women are getting sufficient dietary fibre.[80]

But getting enough needn't be hard. In the table below, I've provided some convenient tips to help you boost your fibre intake. One food group that's particularly rich in fibre is wholegrains: worthy of a special mention because studies have shown that a good intake of wholegrains is associated with a lower NCD risk.[81] Of course, we can't simply point to the rich and varied fibre content of wholegrains to explain the benefits of these gems: wholegrains are also packed with vitamins, minerals and phytonutrients, and are a useful source of protein and slow-release carbs. More likely, it's the combination of all these attributes that means making wholegrains a regular dietary staple a wise choice.

You may also have heard about the benefits of the Mediterranean diet. This style of eating is based on the traditional foods routinely enjoyed by people from southern European regions. These populations typically live long and active lives free from chronic disease. This style of eating is rich in fruit, vegetables, olive oil, wholegrains, beans, nuts and oily fish, and research has shown that those following Mediterranean diets typically have lower incidence of NCD and improved longevity.[82] However, again, it's likely a combination of factors that makes this way of eating good for our health, as people who adhere to the Mediterranean lifestyle are also typically very active, incredibly social and community-focused.

Easy Ways to Boost Your Fibre Intake

- Have plenty of veg with your meals, either cooked or as a salad.
- Bulk out your meals with pulses, like beans and lentils.
- Choose wholemeal or granary bread, wholewheat pasta and brown rice.
- Opt for higher-fibre breakfast cereals like Weetabix, Shredded Wheat, porridge or muesli.
- Add berries to salads and cereals.
- Have at least two items of fruit per day.
- Eat your potatoes, carrots, apples, pears and peaches with their skins on.
- Nuts and seeds are packed with fibre and make great snacks.
- Chia seeds and flaxseed can be added to salads, cereals and smoothies.

- Other fibre-rich snack ideas include raw veg batons, oatcakes and hummus, dried fruit and avocado.
- Fibre acts like a sponge, so make sure you also drink plenty of fluid.

Marketing for Performance

It was back in 1988 when I entered a public gym for my first training session. In the 1980s, the primary function of a gym counter was to take membership fees, sell branded merchandise and maybe serve the occasional cup of water, tea or coffee. Now, visit any commercial gym and not only will you observe a blender ready for the staff to make you any one of several great-tasting protein, weight-gain or meal-replacement shakes, but you'll also likely see a fridge displaying a range of sports drinks and a shop well stocked with numerous tubs and pouches of powders, capsules and other 'performance-enhancing' tonics. The topic of sports and performance nutrition is full of contradictory information, particularly so when it comes to muscle-enhancing supplements.

For some, exercise is seen as a necessary evil: we know it's good for us, we don't particularly enjoy it, but we do it nonetheless. For others, sport is not only an enjoyable pastime, but something that's done competitively. However you view it, physical activity supports positive wellbeing, and anything that helps us to improve our performance and maximise our ability is something that should be nurtured. Contemplative nutrition involves optimising the science-led health benefits available to us, and a good diet – along with exercising correctly and getting sufficient rest – is integral to progress. The debate is not whether good nutrition helps us to improve our performance in sport – there's no doubt that it does; rather, it's about what good nutrition for optimal performance looks like. There's a good reason why you don't see many professional athletes gorging on junk food during their time off or drinking the latest sugar-loaded soft drink. In July 2021, Cristiano Ronaldo, the Portuguese-born footballer renowned for his extraordinary work ethic and jaw-dropping physique, came under intense criticism for moving two Coca-Cola bottles aside at a Euro 2020 press conference, urging onlookers to drink '*agua*' ('water') instead. This move sent the internet into a frenzy, enraging thousands of football fans and sending the giant food company's share price tumbling within a few hours.[83] Structured eating practices are becoming the norm in professional sports, where success or failure depends heavily on their body's performance. But the quest to eat only the 'right' food is not unique to top-level sports. Consumers all over the world are beginning to change their eating habits with a new yearning – rightly or wrongly – for reading labels, understanding where food comes from and what to avoid. You would, however, be wise to question many of the claims about performance nutrition in bodybuilding and fitness magazines and websites. Navigating the plausibility of marketing claims surrounding the multitude of nutritional supplements can seem daunting, especially as there are ergogenic aids where the benefits are well validated, such as for creatine (a supplement that helps to maintain a supply of energy to working muscles[84]) and protein powders (discussed below). But many performance-enhancing supplements offer little benefit despite the citing of scientific references as part of their tactical marketing.

Although I won't be discussing performance nutrition in any great depth, I've shared a few fundamental tips below. As well as these suggestions, be mindful of misinformation and heed the watch-outs on page 44. Always have your rejection detector set to 'high'. Good nutrition for sports performance is much like good nutrition for health: eat good-quality food including plenty of fibrous, carbohydrate- and protein-rich foods, and don't forget your essential fats. Additionally, adjust your diet to suit your energy requirements factoring in whether your goals involve weight control. You may also benefit from adjusting the timing of your meals to help ensure that you're adequately fuelled for hard exercise sessions without eating too much immediately prior to a workout.

Tips for Optimal Performance Nutrition

- Stay hydrated. Not only is it useful to sip water throughout a workout, make sure you're drinking sufficient amounts before and after training, too.
- Get plenty of protein. Include high-quality protein sources throughout the day, such as beans, lentils, tofu, nuts, eggs, fish and lean meat.
- Consider carbs. While not essential, fibre-rich carbs, such as wholegrains, can be valuable for providing energy.
- Essential fats. Fats are energy-dense and vital sources of energy.
- Fibre. Fibre aids digestion and is fuel for the microbiome which, in turn, assists performance.
- Timing is everything. Have a structured plan to provide the right balance of nutrients to fuel and recover from exercise.
- Balanced eating. As well as food sources of proteins, carbs and fats, include plenty of fruit and veggies for fibre and micronutrients.
- Supplements supplement. While some supplements, like protein powders, can be extremely useful, they shouldn't replace good food. And watch out for outrageous claims that most supplements make.

- Eat, monitor and adjust. Be prepared to flex what you eat according to how you perform and feel.

Wheying In

Protein powders have an interesting history. When I first became interested in building my muscles, the only ones available were gritty, foul-tasting milk and egg-white powders. Fast forward just a few years and everyone was talking about whey protein.

The first protein supplements for gym-goers were created in the early 1950s by Irvin Johnson, a self-taught nutritionist who later changed his name to Rheo H. Blair, and Bob Hoffman, owner of York Barbell and coach to the United States weightlifting team from the late 1940s to the early 1960s. They launched Johnson's Hi Protein Food and Hi-Proteen, respectively.[85] Although whey was initially discovered when the first cheeses were produced over 7,000 years ago,[86] it wasn't until 1936 when a pharmacist called Eugene Schiff developed a method of processing whey from milk for human consumption.[87] 'Schiff Bio-Foods' sold what was effectively whey protein isolate with little commercial success. The protein supplement boom didn't really happen until the 1990s when 'wheymania' hit the gym scene.

Nowadays, our choice of protein powder is a lot broader than egg white or whey powders. Plant protein powders – derived from sources like pea, soya, rice and hemp – are the latest to explode, with this 'niche' market being worth over $5 billion in 2020.[88] As the protein supplement market as a whole continues to boom, the latest predictions are that it will be worth over $36 billion by 2028.[89] And it's not just protein powders aimed at bodybuilders that have caused this explosion. The shelves of fitness websites are piled high with a huge array of protein-enhanced drinks, bars, pancakes, cereals, bread, muffins and cookies – even nachos and croissants are being fortified with protein! All these products are being sold, not just to gym-goers, but to the masses with the marketing-led inference that additional protein is good for your health or will enhance gains. According to the market research firm Mintel, 6.1 per cent of food and drink product launches in 2020 claimed to be high-protein or contain added protein, up from 3.3 per cent in 2016.[90]

But clever marketing spin around protein is nothing new. Dubious marketing claims about protein supplements have been there from the outset with Hoffman's Hi-Proteen in 1952. For example, in his pamphlets, Hoffman would write things like 'The production of a "miracle food," such as High-Proteen, is not a hit-or-miss affair. A world famous food research laboratory is put to work. Their chemists and the doctors, who are a part of their organization, work out the product.' Scientific-sounding claims were being made more than 70 years ago: 'After a lengthy period of research and testing, the proper blend is obtained. It must be nutritious, containing – as far as possible – all the necessary amino acids, and it must be pleasant to the taste, so that using it is a pleasure.' Hoffman ended one of his sales spiels with: 'We never leave anything to chance.'[91]

Do we really need all this additional protein? The research does, indeed, indicate that protein requirements for athletes are higher than for the general population,[92] who benefit from extra to support muscle growth and repair. The BNF states that strength and endurance athletes require 1.2 g to 2.0 g of protein per kilogram of body weight per day, compared to 0.75 g/kg for the general population.[93] They note, however, that most diets already contain considerably more than 0.75 g/kg. The BNF also points out that athletes might benefit from consuming 15 to 25 g of protein 30 minutes to two hours after a workout. As most athletes consume a larger quantity of food than non-trainers, an 80 kg male athlete consuming 2,500 to 3,000 calories a day would likely already be gleaning 160 g from his diet. Protein supplements can, however, make consuming adequate protein considerably easier especially if you're controlling your calorie intake, and protein is more satiating than the other macronutrients.[94] Many people can benefit from moderate supplementation with a good-quality protein powder.

Research Manipulation

One bodybuilding supplement that was popular in the early 2000s was HMB (β-hydroxy β-methylbutyrate). HMB is a metabolite of leucine, an essential amino acid with a key role in the initiation of protein synthesis, and athletes consuming plenty of protein will have a good intake of leucine at each meal.[95] Research has shown that HMB can help to prevent muscle wasting[96] and, due to this, it's marketed as an 'anti-catabolic agent'. If you can prevent muscle from breaking down, the net benefit will be bigger muscles, right? Well, yeah … but this is not how HMB works. The benefits of HMB supplementation have been shown in patients where the illness involves a significant amount of muscle atrophy like in cancer cachexia or patients with AIDS.[97] The benefits of HMB to athletes and bodybuilders are dubious and inconclusive.[98]

HMB is no more interesting than any of the multitude of other tonics out there, but it shows us how research is easily manipulated to promote a supplement. When I was working in the bodybuilding scene, HMB was extremely popular: every fitness magazine displayed several adverts for it and it was covered on thousands of bodybuilding websites. HMB illustrates what holds true for most sports supplements today: their marketing includes scientific claims and it's possible that some may have small beneficial effects. For many, there is absolutely no credible evidence – like the arginine pyroglutamate plus lysine that my teenage self gobbled down. HMB is an example of a supplement which, through the manipulation of scientific studies, created much excitement based solely on the cherry-picking of data for, at best, wishy-washy credibility. The sports-supplement industry is rife with this kind of behaviour.

Boosting Your Workout

In the stressful 21st century, exercise is something that often requires us to conjure up extra energy from nowhere. After a hard day's work, the most motivated among us might be able to arrive at the gym, but even they will often struggle to get sufficiently psyched-up for a hard session. So it's understandable why people want a little boost. And for many, a cup of coffee isn't enough. Energy drinks warrant a mention because of their prevalence: the global energy drink market is already worth more than $86 billion and is predicted to reach $150 billion by 2030.[99] But it's not just gym-goers who are consuming them; in fact, it's non-exercisers who are guzzling more of them. In fact, they aren't really 'energy drinks' as their calorie content is generally no greater than other sugary soft drinks and sugar-free versions are readily available. A better term would be 'stimulant drinks', as stimulants are precisely what the active ingredients, like caffeine, are. But, as the term has negative connotations, they're more commonly referred to as 'energy drinks'.

Pre-workout formulas are available in powder, drink and pill form and typically contain a concoction of several ingredients in varying amounts designed to help boost energy levels for exercise. Anyone who's taken a strong pre-workout won't need convincing scientific literature to validate their effectiveness. There's absolutely no doubt, pre-workout supplements 'work'. Enter the gym lethargic, neck one of these ergogenic aids and in 10 minutes you'll be exploding through the most gruelling of workouts. It's claimed that some pre-workouts might even make you feel like you're 'off-your-head' as the effects can be drug-like! These effects make it easy to market these supplements: simply add a dose of key ingredients and the boost is almost immediate.

But are the ingredients safe? Common ones like creatine, beta-alanine, citrulline malate, caffeine, L-theanine and taurine are well researched, and, at the levels stated in the studies and when consumed on their own, they're considered safe. However, in the super-normal quantities and in the combinations that some concoctions contain, their safety is less well established. There's little, if any, research on regularly consuming multiple 'boosting' ingredients in combo. Although some common pre-workout ingredients, like nitric oxide (designed to give 'muscle pumps'), are safe, users complain of side effects like diarrhoea and stomach upsets. More concerning are ingredients like DMAA (1,3-dimethylamylamine), a stimulant designed to give a quick spike of energy. DMAA was banned following the death of a 30-year-old marathon runner in 2012 after she consumed a popular pre-workout supplement that contained it.[100] When taken along with other stimulants, such as caffeine, and under conditions of extreme exertion – i.e. the very circumstances it's designed for – DMAA can raise blood pressure and lead to cardiovascular problems ranging from shortness of breath and tightening in the chest to heart attack.[101] Even caffeine, the most widely consumed and researched psychoactive compound, isn't without its risks. Although caffeine is considered safe if consumed at levels less than 400 mg per day (energy drinks typically contain 100–200 mg per serving), higher amounts over long periods are ill-advised.[102]

Contemplating Food, Nutrition and Health

The key takeaway from this chapter is that any connection between diet and health is extremely complex. Placing too much emphasis on linking a particular nutrient or food to disease risk or to improved health or performance has limited validity. Focusing on a single nutrient factor should not be the goal of dietary advice or nutrition policy. Nutrition science doesn't work this way. Any approach to the way we view our food should not be overly reductive. Moreover – as Joshua Wolrich emphasises – people should be able to eat more or less whatever they like without fear of judgement, especially if the judgement comes from those whose opinion is born out of limited evidence.

So, we have limited reliability when it comes to food and nutrition policy because some of the recommendations are based on insufficient data and outdated research. But should we be completely dismissing it? And, more importantly, which foods and nutrients should we be eating more of and which should we be steering away from? Food policy does, nevertheless, provide sensible recommendations and there are several areas where there is consensus among nutrition experts, with some fundamental tips serving as sound dietary advice.

Firstly, eat a varied range of foods as this helps to ensure that you're obtaining good intakes of essential and beneficial nutrients. Next, macronutrients: there's no justification for viewing either carbs or fats as 'bad' despite what the dietology camps tell us. Thankfully, there's complete agreement when it comes to trans fats: avoid them, which, fortunately, is easy these days. Include sources of protein at each meal such as pulses, legumes, eggs, fish or lean meat. Consume plentiful amounts of fibre-rich foods – fruit, vegetables, pulses, wholewheat grains, nuts and seeds – foods that are also rich in the vitamins, minerals and phytonutrients that we need. There's broad consensus that consuming too much added sugar is ill-advised. Note that I say 'added sugar': fruits, for instance, are nutritious gems, rich in fibre and micronutrients, despite being reasonably high in naturally-occurring sugars. Our ancestors have been consuming fruit for millions of years. Astoundingly, I've come across questionable 'nutrition experts', such as carnivore and keto diet advocates, who suggest that people cut down on carrots because – according to them – they 'contain a lot of sugar'.

The food and botanical journalist Michael Pollan provides a sensible way of viewing food in his short book *Food Rules: An Eater's Manual*. He recommends a more food-first way of eating, rather than focusing on how much of a particular nutrient we consume. He also suggests moderating the amount we eat: 'Nowadays we think it is normal and right to eat until you are full, but many cultures specifically advise stopping well before that point is reached'.[103] When navigating confusing food-related decisions, be mindful of dietary and health claims. Remember, eating one meal that's high in sugar and low in fibre won't ruin a great diet. Nor will taking a supplement fix a bad diet. In short: when it comes to nutrition advice, keep your bullshit detector set to 'high'.

Contemplative Food, Nutrition and Physical Health Key Points

- Make food choices that prioritise your own health.
- Plan ahead, especially if you exercise regularly.
- Eat a wide variety of foods.
- Consume plenty of fibre-rich foods.
- Carbs are good; fat is good – just watch out for added sugar and trans fats.

- Include omega-3-rich foods.
- Have sources of protein at each meal such as pulses, legumes, eggs, fish and lean meat.
- Be cautious about nutrition and health marketing claims.
- Be extra-cautious when it comes to dietary advice from influencers and media personalities. Seek out experienced professionals while keeping in mind they too will have their biases.

As humans have the ability to acquire nutrition from a wide variety of foods – and we're able to process nutrients from a wide range of foods – your diet should focus on selecting foods that support optimal health, both for yourself and for others. Contemplative nutrition involves looking after your own health first: take care of yourself before others. If your own health is poor, how can you care for others? Eating to maximise your own wellbeing will help to keep you out of hospital and means that you'll utilise fewer resources that can then be better spent elsewhere. Looking after your own health means more than just taking care of your body, however: mental, cognitive and emotional health are important too. Although the science linking what we eat to mental wellbeing is very much in its infancy, there are some really interesting areas worth looking at.

Chapter 5

Contemplating Food and Our Wellbeing

As far back as the Stone Age, those who behaved in ways that weren't 'normal' were said to have succumbed to the work of evil forces like demons, gods or witches. Even then, 'experts' were claiming there was a link between human behaviour and what people consumed, and their patients were treated with a prescription of herbs, the ingestion of foul-tasting drinks or starvation.[1]

These days, thankfully, we live in a period where mental health is increasingly being taken seriously. But even as recently as 20 years ago someone who was depressed or anxious was often accused of being weak-minded or was confronted with questions like, 'What have you got to be down about?' Although the legacy of these previous perceptions lives on, societal views on mental and emotional health seem to be moving in the right direction. It's refreshing to know that taking care of our mental health is now part and parcel of many people's lives. You'll hear people claim that exercising regularly helps them mentally and more and more people routinely practise mindfulness, keep a gratitude log, meditate or consult a therapist. However, considering its overwhelming importance, it's perhaps surprising that food is seldom associated with how we feel and that the influence of diet on our wellbeing is often overlooked. By comparison, the powerful effects of food on our physical health are routinely acknowledged as if a nutritious diet only affects our bodies and not our minds.

Mental disorders can be defined as a 'clinically significant distur-
bance in an individual's cognition, emotional regulation, or
behaviour'.[2] This broad definition includes depression, anxiety,
bipolar disorder, personality disorders, psychosis and more. However,
disturbances to mental wellbeing needn't be limited to classical
diagnosed disorders. Traumatic events, day-to-day news, petty
annoyances and even the unexplained 'blues' can all affect people's
mood and add up to long-term issues.

Disorders of mental health are one of the leading causes of illness
and disability worldwide. In 2017, a landmark *Lancet* report that
explored the global burden of disease revealed that over 970 million
people live with a mental health disorder – that's more than one in
eight people globally.[3] Of this, 284 million people experienced
anxiety disorder in the space of one year – making it the most prevalent mental health disorder. Depression comes in a close second
place, with 264 million people directly affected. Even more unsettling
is that more than 700,000 people die to suicide every year, over
three-quarters of whom are in low- and middle-income countries.[4]
For comparison: the UK has 3.7 times more suicides than road fatal-
ities.[5] These figures don't account for the many more individuals
whose mental state leads them to attempt suicide. Countries report
mental health differently and various cultures have contrasting
perceptions of mental health, which means that there is a huge
disparity with diagnosis and data between nations. Consequently, it's
likely that the above figures are under-reported, and it's estimated
that more than two-thirds of people either don't seek help or fail to
receive support.[6] Add to this the fact that it's not only those who are
directly afflicted who struggle: their long-suffering loved ones bear
much of the burden.

Feeding Your Inner Self

The role of diet on mental health is far from fully understood. For all the talk about finding your inner self in the depths of a freshly harvested apple, messages about nutrition have been contradictory and changeable. What this has led to is a situation where the sparse evidence available is constantly placed under intense scrutiny and often misrepresented by social media influencers. This has limited progress in embracing the benefits of nutrition and fostering scepticism around the quality of scientific work in this area. In the hope that this section will help you to find some clarity in this mishmash of opinions, first, remember the learnings from the previous chapter: good nutrition is essential to your health, and there's increased awareness of the well-established link between diet and NCDs like type 2 diabetes, heart disease and some cancers. However, there are encouraging reasons to believe that this burgeoning trend is also, in part, fuelled by a growing understanding that physical health is intimately linked to your mental health.

Perhaps the strongest evidence of this comes from conditions with an obvious link to diet. Eating disorders like anorexia nervosa, bulimia and compulsive eating involve food being used as a means of control, a manipulation tool or a way to affect mood. Certain genetic conditions involving defects in the way specific nutrients are metabolised have consequences for mental development, too. These are known as inborn errors of metabolism and an example is phenylketonuria (PKU). Newborns with PKU lack an enzyme and this causes levels of dietary phenylalanine (an essential amino acid) to build up leading to impaired brain development.[7] Consequently, infants have to adhere to a diet where their protein intake is strictly controlled. A less serious genetic condition where cognition is impaired is haemochromatosis where the body stores too much iron. Those affected have to moderate their intake of dietary iron to avoid levels building up, otherwise they risk suffering from excessive fatigue and struggle to concentrate.[8]

Beyond eating disorders, the link between what we eat and how we feel is similarly obvious. Take, for instance, that post-prandial afternoon slump: you're staring at a computer screen struggling to focus and you look around the office to see if anyone will notice you cradling your head in your arms to grab a few minutes 'shut-eye'. I bet many of us have experienced a low mood for a day or two after a heavy night's drinking. Then there's the modern colloquialism 'I'm hangry!' announced by someone casually claiming to be in a crappy mood due to a lack of sustenance. Although it might be questionable that we have a genuine dire need for food, the term provides an apt acknowledgement that we're aware that food – or lack of it – profoundly affects our mental state. The link between food and mood is glaringly apparent: we associate food with enjoyment. The pleasure we derive from eating our favourite meal makes us feel good. When we feel low, we open the kitchen cupboard and grab a tasty snack. It's not only when we're depressed either: we munch on tasty snacks to further promote the hedonistic buzz while we're enjoying other activities, like chomping on popcorn to boost the pleasure from watching a movie. However, what you choose to eat may be limited to a mere short-lived mood lift if you make poor choices. Over time, junk-food diets high in sugar and low in key nutrients may have devastating consequences for your mood.

Dieting is Mentally Draining

Even if you've never attempted to cut back on what you eat to shed a few pounds, you'll know people who have. Dieting to lose weight is the most commonly followed type of dietary manipulation[9] and, in 2020, four in 10 Americans followed such a regimen with calorie counting being the prevailing approach.[10] There's good reason why people watch how much they eat: in the last chapter, we saw how being a healthy weight can reduce the risk of chronic disease despite the controversy about what's meant by a 'healthy weight' (when defined by the flawed BMI) and the process of weight control being extremely complicated. Because of the negative stigma surrounding body image, society's judgemental attitude about appearance and the continual pressure to lose weight, too often dieting may not be the best approach, either for our physical health or our mental wellbeing.

Being on a 'diet' impacts on our mental state especially if we drop our calories too much. Restricted energy intake, in turn, restricts our brain power. When dieting, we have slower reaction times and a reduced ability to concentrate.[11] Diets make us feel isolated, lower our mood and create such a preoccupation with food that all we want to do is to eat. All of the stress and anxiety about food and weight that preoccupy dieters can consume a portion of their working memory,[12] and chronic dieting can lead to feelings of depression, low self-esteem and increased stress. Often this is because dieters set unrealistic goals, which, when not achieved, result in feelings of failure. There's an internal war: the overwhelming biological impulses versus the pressure to be a desired weight. The result: reduced resistance, increased pleasure from eating and an obsession with food. Food becomes the most important thing in our life and this triggers further eating. These impulses can be especially hard to combat in the modern world when food cues are all around us. Hunger chemicals rampage our brains sending our mood all over the place, and all we hear is food noise. Although dieting may not cause an eating disorder, the constant concern about body shape, fat grams and calories can start a vicious cycle of body dissatisfaction and obsession that can all too quickly lead to one. We're controlled by our cravings.

Encouragingly, over time, the impulse to reach for our favourite foods can be reduced. Although dieting in the short term may mean we're more likely to grab a snack, by taking a sensible approach to long-term controlled eating, it's possible to quash these conditioned food cravings simply by consuming the foods we crave less often.[13] When we get used to not having our favourite indulgences, that suits us just fine. However, this involves willpower, which is something that comes more readily to some than others.

Focused Eating

We've all been there. The end of a long and stressful day. We're tired, hungry, moody and struggling to concentrate. Lack of food means a lack of energy and the mere thought of having to appear full of vigour after a long drive home drains that last modicum of energy. When our blood glucose level is low, not only do we feel physically fatigued, but we struggle to concentrate, and our willpower and self-control are impaired.[14] This is why when you're hungry and unproductively trying to focus on finishing that urgent assignment for your boss, the first thing you should do is to eat. Research has shown that when we haven't eaten for a while, a single act of self-control causes blood glucose to drop further, making it harder to maintain self-control on another task unless we eat.[15] For us to stay motivated, efficient and focused on multiple things, we need to feed ourselves regularly. This may be why people trying to quit cigarettes find they crave sweet foods.

When it comes to nutrition, our mood and ability to concentrate aren't only controlled by how much we've eaten. Our mental state is also affected by specific nutrients and the overall quality of our diet. Give a six-year-old a bag of sweets and there's little hope he'll be sitting still 10 minutes later ready for you to read him a story; more likely he'll be running around pretending to be Spider-Man. Sugar doesn't just cause hyperactive kids, either: diets habitually high in refined carbs impair cognitive function in grown-ups, too.[16] Gobbling sugary snacks affects our short-term concentration. A 2018 New Zealand study compared the cognitive effects of the dietary sugars glucose, fructose and sucrose to placebo (in the form of the sweetener sucralose). Subjects were given a drink containing one of the sugars, and the group was also divided into two groups, fasting and non-fasting, in order to observe the effects in fasted states. The subjects were given cognitive tasks – which assessed response time, arithmetic processing and attention – and the results showed a negative effect on cognitive functioning from glucose and sucrose, with fructose having no effect. The effect was notably more pronounced in the fasting group. According to the authors 'the "sugar coma" – with regards to glucose – is indeed a real phenomenon, where levels of attention seem to decline after consumption of glucose-containing sugar'.[17] Maybe that bag of Haribos on your desk isn't such a good idea, after all.

As well as the simple carbs we devour, our cognitive function and wellbeing are influenced by how regularly we feed ourselves, the nutrient composition of our meals, our background nutritional state, eating habits, food-related beliefs and the nature of the mental tasks we're performing.[18] For example, having a good breakfast is associated with improved cognitive performance later in the morning, while the wrong lunch choice can lead to the dreaded post-prandial slump, and a healthy meal later in the day appears to boost our ability to perform tasks involving sustained attention or memory.[19] Things are less clear cut when it comes to the effects of fasting. Experiments that have explored the effects of intermittent fasting – i.e. abstinence from food for more than 12 (often 16, 18 or more) hours for reasons concerning health, religion, discipline, employment or personal preference – on our ability to concentrate, have provided varied and nuanced results. For instance, research on sleep-deprived night-shift workers has shown that eating too large a meal may reduce the ability to perform, and smaller snacks are preferable.[20] Studies on Ramadan-fasting Muslims have indicated a negative impact on educational performance, mental health and decision-making strategies.[21] Others contradict, showing little adverse effect on levels of cognition.[22] This huge variability in results was demonstrated when a 2021 meta-analysis on temporary abstinence from eating and its effects on cognitive function failed to draw firm conclusions.[23]

Does this mean that fasting will help you stay glued to your monitor for hours or will it send you drifting into daydream and procrastination? Simply put, it's hard to say whether the best strategy to stay focused is to chomp down three square meals a day, graze on frequent snacks or to skip breakfast altogether. Most likely, multiple factors come into play and, if your overall diet is low in sugar and rich in quality food, maybe it doesn't matter. Although forced abstinence from food may help boost mindset and resilience, when it comes to the issue of fasting and wellbeing, this may be one of those rare occasions when scientists are forced to admit that subjective feelings trump objective data. Deciding if you feel better from prolonged fasting or sticking to regular meals might well be down to you making up your own mind. What's optimum for someone else might not be the best strategy for you. There is, however, one group of people where the research into the benefits of fasting on cognitive health is extremely encouraging. Several studies using intermittent fasting as an intervention on elderly subjects have indicated improved outcomes with possible benefits for those suffering from brain-related disorders such as dementia, Parkinson's and stroke.[24]

There's one class of nutrients where the evidence is a lot clearer. Research has shown a strong link between omega-3 fatty acids and cognition. In Chapter 4, we saw that modern diets have lower amounts of omega-3s with dire consequences for our health. DHA in particular seems to play a key role in ensuring our cognitive function is maintained as we age.[25] This semi-essential fatty acid seems to protect against age-related cognitive decline (ARCD), Alzheimer's disease and other brain disorders,[26] while enhancing memory, reaction time and strengthening neuroprotection in general.[27] Does this mean that DHA might be our silver bullet in our fight against dementia? While it's unlikely that DHA is the brain's panacea for ARCD, there's reasonable evidence that supplementation with DHA goes some way to improving learning and memory as we grow older.[28] DHA's fish-oil partner-in-crime EPA has also been linked to cognitive function. EPA might help to support attention deficit hyperactivity disorder (ADHD) in children. One encouraging study showed that kids who consumed diets low in omega-3s benefited from fish-oil supplements to a level that was at least as effective as conventional pharmacological treatments.[29] Non-fish eaters needn't despair, however, as algae-derived omega-3 supplements are available and diets rich in plant sources of ALA can allow for sufficient conversion to DHA and EPA.

A Hug From a Mug

Maybe all that we need is the feel-good factor of a hot drink to perk ourselves up. Coffee is one of the most consumed beverages in the world with around two billion cups being imbibed every day[30] – that's about 83 million per hour. This dark, rich liquid began its global expansion from the highlands of Ethiopia where the stimulating effects of caffeine were first observed.[31] Today, Americano cups populate cities across the world and most people will admit to not being able to function without first downing their double-shot espresso. So why is coffee so popular and what's driving this insatiable thirst? Its widespread popularity owes a lot to the recognition that coffee is more than a caffeine kick. The main biologically active ingredients are caffeine (which is also found in tea, chocolate and certain other plants) and a suite of antioxidants. Upon ingestion, caffeine is absorbed, metabolised and quickly shuttled to the brain, where it works as a stimulant, at least in part, by blocking adenosine – a chemical that promotes sleep – from binding to its receptor.[32] Caffeine is a psychoactive compound: it changes our mental state by acting on the central nervous system. At low doses (1–3 cups per day), coffee sharpens people's minds and makes it possible for sleep-deprived, underprepared students to pull an all-nighter to meet an impending deadline. A 2021 pan-European survey of over 5,000 individuals found that consuming 75 mg of caffeine – the equivalent of one small-to-average cup of coffee – every four hours, led to substantial mood improvement over the course of the day.[33] This finding fits well with at least one previous study, conducted at the Harvard School of Public Health, that indicated that caffeine might even work as a mild antidepressant, lowering the risk of depression by as much as 20 per cent in women.[34] But there's a catch. Overuse – over 500–600 mg a day – can lead to 'caffeine intoxication': unpleasant side effects that include wrecked sleep, shakiness, irritability and gastric distress. Excessive reliance on caffeine can further turn dependency into a hard-to-shake habit, leading to a full-blown substance-use condition that's been named 'caffeine-use disorder', listed in the *DSM-5*,* and can lead to pervasive mental health conse-quences.[35] Caffeine, it seems, doesn't just keep you alert but has the power to turn you into a walking ray of sunshine, provided you don't overdo it.

This leads us to another important aspect of coffee and, indeed, other beverages: temperature. For most people temperature is the reason why tea and coffee have become a morning ritual: the indulg-ing heat of our morning cup helps fight off the foggy haziness that accompanies us out of bed. In fact, hot beverages can also work the

---

* The *Diagnostic and Statistical Manual of Mental Disorders* (*DSM-5*) is the fifth edition of the standard classification of psychiatric diagnosed and mental health disorders.

other way: if you're feeling tired or under the weather, it's comforting to climb into bed with a hot mug in your hands. The temperature of our drink seems to have a direct effect on our mood and cognition. But does it, really? The neuroscience of thermal sensation is fraught with examples of how rises in temperature evoke positive emotional responses in humans. In a 2017 study, participants were asked to rate either coffee or green tea served at cold (5°C), ambient (25°C), and hot (65°C) temperatures. Not surprisingly, beverages consumed at 65°C were more likely to trigger 'emotions of positive valence, such as 'pleased', 'happy', 'warm' than those consumed at either 25 or 5°C'.[36] Not only do hot drinks make people happy but they can also sway our decision-making. Research shows that humans rate strangers as 11 per cent 'warmer' when holding a hot drink in their hands compared to holding a cold drink.[37] Although some psychologists feel it's a contentious claim, the warmth of a drink might determine how friendly you are towards someone you've just met. While interesting, this falls short of explaining other behaviours linked to hot drinks. Why do we offer a hot cuppa to an upset friend? In Britain, the gesture of offering tea is presumably rooted in tradition: sharing a cup of tea is an act of compassion and offers an effective social support strategy. In the face of its cultural pervasiveness, it might perhaps then come as a surprise that scientists are only just beginning to untangle the benefits of drinking tea on mood and cognition. In 2014, for example, researchers reported that in a healthy Japanese population, those who drank four or more cups of green tea per day were 51 per cent less likely to develop depression over their lifetime than those who drank less or no tea.[38] But, as we have learned, epidemiological studies have limitations and other factors are equally likely to be responsible for the positive results, such as diet, socioeconomic status or, simply, genetics.

The take-home message? If you drink coffee in moderation, savour those sips, harness the surge in energy and enjoy the benefits. Too much caffeine, however – especially too late in the day – inhibits sleep, and having too little sleep, of course, will more than cancel out any caffeine-derived cognitive benefits. We don't yet know how coffee can alleviate symptoms of depression, but it may be to do with its special combination of antioxidants that are thought to dampen pathological changes in the brain of depressed individuals. For instance, polyphenols with anti-inflammatory potential may protect against low-grade inflammation in depression.[39] Another hypothesis is that caffeine facilitates the release of feel-happy chemicals in the brain such as serotonin and dopamine.[40] It's suggested that caffeine stirs up the amount of dopamine that binds to dopamine receptors in the forebrain – a key area of the brain involved in motivation, decision-making and reward. When it comes to the science of hot drinks, we'll need to wait a fair few years before fully appreciating their effects on our wellbeing. Remarkably little is known about how tea can help us to escape the clutches of depression. Although there's convincing evidence on improvements to mood in healthy populations, researchers are still trying to identify the active compounds that give tea its mental health benefits. Despite there being little evidence – at least for now – showing that tea can help improve our mood, in a world oppressed by ever-increasing rates of mental health struggles, it may have far-reaching consequences. Besides, who doesn't love a good old cuppa?

Gut Feelings

In recent years, large studies have provided evidence that nutrition may play an important role in the prevention, management and recovery of psychological disorders. Depression provides a gleaming example. Research into the role of nutrition on depression has consistently shown that good adherence to traditional diets – such as the Mediterranean style of eating – is associated with a reduced risk of depression compared to the less than idyllic Western diet.[*][41] Scientists seem to agree that this difference is, at least in part, determined by the fact that traditional diets tend to be high in fruits, vegetables, nuts and legumes with moderate consumption of poultry, eggs, dairy products and red meat. Such diets are also virtually devoid of highly processed and refined foods, of which the Western dietary pattern abounds. Of course, this is a rough generalisation: there's a great variation of foods eaten around the Mediterranean Sea – let alone between traditional diets around the world – and other factors are also at play. Many of the tasty and calorie-laden foods have the overall effect of increasing your blood glucose, sending your body on a rollercoaster of high-to-low glucose cycles. These foods can cause such a sharp spike in blood sugar (the 'high'; hyperglycaemia) that your pancreas must secrete more insulin to bring your blood sugar back down. Sometimes, the pancreas overshoots and brings the blood sugar down too low (the 'low'; hypoglycaemia). The dizzyingly rapid rise and fall of blood glucose – and resulting compensatory mechanisms – sets off alarm bells, and your body responds by secreting autonomic counter-regulatory stress hormones such as cortisol and adrenaline. Clinical studies conducted under laboratory conditions have shown that these signals increase the likelihood of experiencing anxiety, irritability and hunger.[42] Moreover, consumption of highly refined carbs has been shown to increase the risk of depression in otherwise healthy individuals.[43] While compelling, another explanation is also plausible. Malnutrition and obesity are closely intertwined with cardiovascular disease, leading some to believe that increased susceptibility to mental disorders could be a simple side effect of other pathologies.[44]

---

* Traditional vs Western diets have been shown to have a 25–35 per cent reduced risk of depression.

The above findings provide a useful framework to further investigate the connection between food and mental health. Healthy food props up your physical health with benefits extending to your emotional and psychological wellbeing. Here, we can see that food is indirectly connected to your behaviour and mood. But what about food having a direct effect on your mental health? What if certain foods could make you more likely to jump out of bed with infectious energy while others make you prefer to stay safely tucked-in? Population-based studies have proved tremendously useful in establishing cause-and-effect relationships between food and mental health, paving the way to the flourishing field of what is now known as 'nutritional psychiatry'. However, these studies typically fail to parse out the exact biological mechanisms that are behind these relationships, the mysterious forces that match certain foods to certain behaviours, moods or even diseases. The brain, of course, has drawn much of the attention in the quest to unveil these mechanisms. This makes sense. The composition, structure and function of the brain are dependent on the availability of crucial nutrients, including fats, amino acids, vitamins and minerals. It is therefore logical that food intake and quality would have an impact on brain function, which makes diet a modifiable variable to target mental health, mood and cognitive performance.

Messages from Tiny Friends

A more recent explanation for a connection between food and mental health has its roots away from the head. In fact, the other end of the alimentary canal is what's making headlines these days: the gut. More precisely, the gut microbiome. The microbiome refers to the long list of some trillion bacteria and other microorganisms that reside symbiotically with us in our guts. These organisms populate the gastrointestinal (GI) tract, rolling in the mucus that lines your bowel.

A unique property of these homegrown bacteria is that they occupy themselves with chewing up nutrients, spitting out their remnants in the form of molecules that pass through the gut wall into the bloodstream. From here, these molecules jostle with numerous other chemicals, toiling with biochemical pathways that take place as far away as the brain.

The microbiome is still poorly understood, but mounting studies suggest that your gut microbiome can affect how much food you stuff into your mouth. These little guys help break down the food we shovel down our necks, facilitating the digestion process and helping our bodies maintain a steady state of equilibrium. And before you think that these tiny organisms only account for an insignificantly small part of your body, consider this: every adult has around a kilogram of these microbes, that's roughly the weight of the human brain.[45] In fact, this enormous mass of microbes inside your gut might have more in common with your brain than you could have ever imagined. Scientists have known for decades that the food we eat changes the balance of microbes in our digestive tract. For example, choosing to eat your favourite yoghurt for breakfast instead of pork sausages and eggs cranks up the amount of some types of microbes while diminishing others. However, what's now becoming clear is that food choice is a two-way street. Research has shown that certain types of gut microbiota – the term that refers to microorganisms residing in a specific habitat, i.e. our gut – can influence how much you eat. The mechanisms aren't entirely clear, but most scientists agree that it's because some microbiota send signals directly to your brain. Specialised cells lining your gut relay information to the brain through unique routes, such as the vagus nerve or the spinal cord. An altered gut microbiome has been linked to obesity and metabolic complications by disrupting energy regulation, inflammatory mechanisms and gut hormones.[46] Essentially, poor dietary choices affect the microbiome which, in turn, affects our appetite and the way our bodies deal with the food we eat: a vicious cycle.

The first hint that these tiny guys could influence your mental state came from a study published in Japan in 2004.[47] In this study, Nobuyuki Sudo and colleagues observed that when certain mice were given a blend of non-pathogenic microbes, their behaviour changed. The mice in question were known as 'germ-free' – they were raised under sterilised conditions so that they don't have microbes in their bodies. Animals raised this way were shown to display aberrantly anti-social behaviours characterised by high levels of stress. Strikingly, soon after ingesting microbes, their stress dropped off sharply suggesting that the gut bacteria were somehow shaping their hormonal response. Although the study caused widespread consternation in the scientific community, the idea that bacteria could alter our thinking processes had been fermenting for a while. In 1910, a young doctor called J. George Porter Phillips made the remarkable discovery that targeting the gut can help those suffering from mental health disorders.[48] In a landmark study conducted at London's notorious Bethlem Royal Hospital, he observed that patients with melancholia – a term previously used to mean depression – often suffered from severe constipation, along with other signs of what he called 'general clogging of the metabolic processes'. Under the assumption that their mental disorder could be somehow connected to their gut dysfunction, he decided to take an orthodox approach and asked his patients to embrace a new diet devoid of all meat, except fish. He also recommended the consumption of kefir – a fermented milk drink similar to yoghurt – which contains the lactobacillus bacteria. The results were striking. Of the 18 patients Phillips tested, 11 completely recovered, with others showing significant improvement.

Following Sudo's observations, other scientists have been inspired to embark on new research spanning the gamut of diet and mental health. In one study, scientists went as far as transferring faeces from depressed people into rats. The paper – cunningly titled 'Transferring the Blues' – represents one of the most convincing and bizarre pieces of evidence supporting the notion that the gut microbiome can alter someone's behaviour and induce depression-like symptoms.[49]

Encouragingly, this exciting area of research is inspiring the development of new treatments. In 2015, Katya Gavrish, a microbiologist trained in Russia, started a company called Holobiome to search for bacteria that may lead to new therapies for depression and other brain disorders.[50] Part of the allure comes from the tacit assumption that certain gut bacteria can alter how the brain works. Gavrish is part of a growing number of researchers thinking that 'psychobiotic therapies' – bacteria-based therapies that promise to treat psychiatric disorders – provide an attractive alternative to conventional drug formulations.[51] Building on the science outlined above, their theory hinges on fairly recent evidence that certain bacteria in the gut can produce neurotransmitters including dopamine, noradrenaline (also known as norepinephrine), GABA and, crucially, serotonin.[52] These chemicals are involved in our mental wellbeing and it's thought that they act as powerful levers that your microbes can pull to influence your emotional state. Once in the bloodstream these compounds can reach remote areas of the brain, leading to profound changes in your mood or your food preferences. This implies that, ultimately, the gut microbiome might have a direct effect on brain chemistry. Scientists have found that these bacteria sit in close proximity to the enteric (gut) nervous system (ENS), which comprises around 100 million neurons in total. Among its many functions, the ENS provides a critical bridge between the gut microbiome and the brain. It does so through the vagus nerve, a long and complex strand of fibres. Researchers have found that the vagus nerve provides the microbiome's secreted products with a fast lane to the brain. Because the ENS is directly linked to the central nervous system (CNS), gut-based bacteria can directly manipulate brain activity through this route. This shows us that activity changes in this region can lead to far-reaching consequences on your mood, anxiety and emotional state, in turn affecting your drive to eat. Feeling anxious may increase the lure of junk food, altering our eating behaviour and, ultimately, increasing the risk of obesity and metabolic diseases.

Other routes in the gut–brain axis exist, including the bloodstream, which provides a back-door to hunger-stimulating hormones. However, there are several reasons why the vagus nerve has been heralded the 'superhighway' of the gut–brain axis. First, experiments in animals have shown that cutting the vagus nerve abolishes several psychobiotic actions.[53] In other words, for all the sway that gut bacteria hold on your brain, there is little they can do without a functioning vagus nerve. Secondly, the gut produces a staggering amount of neurotransmitters. Serotonin, for instance, is produced primarily by the gut, contributing over 90 per cent of total body serotonin.[54] The third reason why the vagus nerve is a quintessentially key element in the gut–brain axis relates to vagotomy – the surgical procedure that cuts off gut–brain communication – and selective serotonin reuptake inhibitors (SSRIs) – drugs that boost the amount of serotonin hanging around in the brain and the most widely prescribed anti-depressants on the planet.[55] Vagotomy extinguishes the depression-dampening effects of SSRIs in animals. This astonishing finding came as a result of laborious work from a team of neuroscientists in Canada who, in 2019, reported their discovery in *Nature Scientific Reports*, highlighting 'the potential for pharmacological approaches to the treatment of mood disorders that focus on vagal stimulation and may not even require therapeutic agents to enter the circulation'.[56] The world's most widely prescribed antidepressants may not be necessary after all, at least according to some scientists. Focusing on the vagus nerve may yield better results. Perhaps this is why since 2005 – the year when this therapy was approved by the US Food and Drug Administration (FDA) – Americans and Europeans suffering from treatment-resistant forms of depression are eligible for a funky treatment called 'vagus nerve stimulation'. This technique involves sending regular, mild pulses of electrical energy to the vagus nerve through a coin-size pacemaker implanted below the skin. Patients treated this way have reported significant improvements,[57] confirming the hypothesis that the vagus nerve – and, by extension, the gut–brain axis – is crucially involved in the mechanisms of depression.

These studies point to two factors that must be considered when thinking about mental health: first, food can directly influence your mental health and, second, it does so – at least in part – through changes to your microbiome. That our brain is intrinsically connected to our gut may come as no surprise to many. Phrases like 'gut feelings' and 'butterflies in my stomach' have been around for decades, and it's well known that messages relating to mental health travel from brain to gut. Under conditions of acute stress, for example, the brain sends signals to the GI tract, leading some people to experience stomach discomfort or bowel dysfunction. However, it's now abundantly clear that the gut–brain axis is bidirectional – signals affecting our emotional and psychological wellbeing also travel from gut to brain. Indeed, the intricacies of this communication system are now attracting the interest of pharmaceutical enterprises keen to develop ground-breaking treatments for psychiatric conditions. In a world ravaged by mental health disorders, this is good news. Existing antidepressants are predominantly brain-centric, altering the balance of key neurotransmitters; however, because not everyone who takes antidepressants shows signs of improvements, many patients popping these pills fail to escape the clutches of depression.[58]

The gut–brain axis offers a promising new frontier of research for more effective therapies against depression and other mental health disorders. We're already finding out that foods that keep your alimentary system healthy will improve your gut health and subsequently your mental health. The microbiome is but one cog in the immensely complex machine that regulates our motivation to eat. Rigorous scientific research in humans is crucial before we can draw definitive conclusions on the extent of the role that our little friends play in our mood and eating behaviour. The next time you're reaching for that chocolate bar that you don't really need, it might be worth considering whether it's just stress talking.

Tips for a Healthy Gut

- Eat plenty of fibre-rich foods, such as pulses, fruit and vegetables (see page 78).
- Have wholegrains and fibre-rich cereals several times each week.
- Include fermented foods such as kefir, sauerkraut and natto.
- Regularly include omega-3-rich foods, such as oily fish, algae products, chia seeds, flaxseeds and walnuts.
- Drink plenty of fluids: fibre acts like a sponge.
- Limit your intake of junk foods.
- Moderate your alcohol consumption.
- Stay active: try to do at least 150 minutes of moderate or 75 minutes of vigorous intensity activity a week.
- Mindfulness practices like meditation or gratitude can help.

Drinking Problems

There's a good reason why alcohol has been enjoyed by multiple human cultures for many millennia: the experiential alteration to someone's mental state from having a drink is generally perceived as a pleasurable one, at least in the short term. Alcohol uplifts mood, boosts confidence, reduces anxious feelings, relaxes us and supports our behaviour in social settings. Yet, despite the fact that social drinking and moderate alcohol consumption can be fun, drinking has a darker side: frequent drinkers experience adverse effects on their cognition, behaviour and mental state. The more people drink the more they become tolerant to alcohol's short-term effects, leading to them needing to drink more in order to mask any mental health issues. Drinking can be a cause or symptom of mental ill health, and can exacerbate conditions like anxiety or depression in those with no previous issues, leading to further drinking as a coping mechanism. Alcohol's demising effects are illustrated by phrases such as 'mother's ruin', attributed to gin during the 1700s epidemic of women hooked on the spirit who mistreated their children; 'Dutch courage', an idiom referring to the excessive bravado advertised in drunk people; and, more shockingly, 'wife beater', unfairly credited to a brand of European lager that used to have a higher alcohol content than its competitors and, it was claimed, was associated with domestic abuse.

Alcohol is a drug, albeit one that society views in a considerably different way to other drugs that affect our mental state. The way we frame alcohol is different. For instance, we don't say we're 'high' when we're intoxicated with alcohol; rather, we use the word 'drunk'. Socially, we may urge someone to enjoy a glass or three of wine after a stressful day – *go on, have a drink; you deserve it!* – yet we may take a dim view of a friend when they recreationally partake in other drugs, even natural ones like cannabis, psychedelic mushrooms or even tobacco. This is partly due to the legal position of alcohol versus other substances. But alcohol's social acceptance does nothing to detract from its negative effects on the wellbeing of society. According to the World Health Organization, 2.6 million deaths every year result from harmful alcohol use: that's around 5 per cent of all deaths globally.[59] In the UK, over 7.5 million people show signs of alcohol dependency according to NHS figures[60] – that's around 11 per cent of the population. Alcohol is associated with antisocial behaviours like assault, domestic abuse, drink-driving, homicide, sexual abuse and self-harm. Lately, things have been looking even grimmer: according to a recent report by the Organisation for Economic Co-operation and Development (OECD) – a club of the wealthiest nations – drinking soared during the COVID-19 pandemic.[61] Current figures suggest that off-trade sales of alcohol in 2020 increased by 25 per cent compared to the year before, leading to more than an additional 12.6 million litres of alcohol purchased in the UK alone.[62]

Although the need to demonstrate the claim that alcohol changes our mental state requires little in the way of scientific study – the proof is evident after a couple of pints – given its ubiquitous use in society and its troubling effects on health, it's worth looking at the biological effects of alcohol consumption. It is no secret that alcohol affects the way our brain works. Both in the short term and over time, alcohol alters the balance of certain hormones and neurotransmitters. In Chapter 2, we saw how any decision taken by our brain – whether that's flinging food down our throat or aimlessly opening the fridge in search for stomach-grumbling inspiration – is ultimately the result of powerful, all-or-nothing excitatory signals. In our quest to understand how our motivation to eat springs up from our subconscious, we have also learned that excitatory signals need to triumph over a vast array of inhibitory signals for any action to be faithfully executed. What does this have to do with alcohol? For a long time, it was believed that alcohol consumption liberates time-strapped, overworked professionals by dampening feelings of anxiety or fear. As most textbooks will attest, alcohol crafts this otherworldly experience by inhibiting circuits of our brain tasked with keeping us vigilant, alert and Excel ready. For instance, having a drink or two enhances GABA,[63] a neurotransmitter involved in calming our nerves and improving focus. But alcohol also suppresses the transmission of excitatory neurotransmitters, such as glutamate, ultimately slowing down brain function so not all information comes through.[64] Combined, these two mechanisms result in increased inhibition and decreased excitation, which helps to explain why people enjoy a relaxing drink after a stressful day's work.

Science has also shown that alcohol's feel-good effects derive from the release of endorphins: 'pleasure' neurotransmitters. These chemicals bind to opiate receptors in the brain, accentuating the hedonic properties of alcohol and potentially increasing its consumption.[65] Experimental evidence indicates that endorphins may affect both the appetitive and consummatory phases of the motivation to consume alcohol, regulating the searching for, and rewarding effects of, alcohol. This is the reason why blocking opioid receptors is used to treat alcoholism. Alcohol also cranks up the levels of dopamine in your brain, tricking you into feeling more confident and relaxed. This usually happens in the early stages of consumption, when most people would see themselves as 'tipsy'. A dangerous aspect of both endorphin and dopamine stimulation by alcohol is that, over time, with more drinking, chronic consumption leads to a blunted effect on both neurotransmitters.[66] This phenomenon helps to explain why some drinkers become addicted: their brains become wired to crave booze when stressed and drink more and more of it to reap its soothing effects.

To make things more complicated, booze seems to take the edge off our rainy days by interacting with two other well-known chemicals: serotonin and cortisol. Most experts agree that alcohol jacks up serotonin levels[67] and in so doing it acts as a social lubricant, lifting your mood and making you feel closer to your friends. When it comes to cortisol, the effects are less clear cut. While alcohol may decrease cortisol levels in the short term, differences in blood alcohol content, family history and study design have recently called this notion into question.[68] One Australian study showed that even the blood sampling procedure can be a significant source of stress in studies investigating the acute effect of alcohol. In other words, your fear of needles might very well cause a sudden spike in cortisol.[69] The bottom line is that at lower levels of intoxication – a blood alcohol level of 0.04 per cent (that's around two drinks for most people) – alcohol is likely to reduce cortisol levels in most people. However, there is no one-size-fits-all formula when it comes to understanding the effect of alcohol on stress management. Accumulating research shows that the effect of acute alcohol on cortisol levels might very much depend on subject-specific factors (i.e. genetic influences, acute stress levels, etc.). Having a drink might make some of us sleepy, some overly euphoric and for others it can cause stress.

But what about in the longer term? Sipping a glass of rosé or having a cold glass of beer can indeed provide you with a bit of relaxation up front, releasing endorphins and boosting serotonin levels. However, after drinking, levels of serotonin drop, and GABA and glutamate have to stabilise, all of which contributes to anxiety and low mood.[70] This so-called 'hangxiety' (a term coined by UK drug expert Professor David Nutt) effect lasts for a day or two, but for people who drink regularly, their mood fails to fully stabilise. When they do stop, it can take weeks to stabilise, enticing them to drink more. Indeed, in alcoholics, Nutt claims to have observed changes in GABA levels years later.[71] It's also been shown that, over time, regular drinking interferes with your body's stress axis, leading to an increase in cortisol levels.[72] The biological basis for this effect remains unclear but, as far as one theory goes, chronic alcohol consumption damages liver function and, therefore, reduces one's ability to metabolise cortisol (which thus remains in circulation for longer).[73] What's worse, researchers have found that high concentrations of cortisol are associated with neurotoxicity, leading to cognitive deficits in chronic drinkers including impairments in the decision-making process, memory, attention and learning abilities.[74] Although it may take a few weeks for most people's mood to improve when they stop drinking, when they've passed this point, abstinence gets easier as the neurotransmitters stabilise.[75]

Contemplating What to Eat and Our Mental Wellbeing

Being told to steer clear of chocolate might be the last thing you want to hear when your mood is low. The texture of this sweet, rich confectionery melting on your tongue tantalises you with a little boost of pleasure: just what's needed when you're feeling a touch of the blues. But recall that the overwhelming impulses that cause us to indulge in our favourite foods is a problem compounded by the modern world of food abundance. As you dive into that third slice of chocolate cake, remember that these signals are no longer working to your advantage, and that this evolved mechanism is allowing us to derive the contentment that we experience from our food.

Does this mean, therefore, that because we're pleasure-seeking creatures in a world of privilege that we're doomed? Considering that we know that the food with which we're tempted is too often destructive for our physical and mental health, how are we to navigate these potentially perilous urges? An especially troublesome quandary when you consider that these impulses are frequently being exploited by food marketers and, it seems, by scientists and journalists, too. When a 2018 survey claimed to have revealed a link between chocolate consumption and the 'reduced odds of clinically relevant depressive symptoms',[76] headline-grabbing journalists descended into a chocolate-touting frenzy. Arguably, this was understandable. After all, scientists were telling us to eat more of one of the world's most enjoyed sweets. What's not to like about that? A finding like this is, itself, enough to lift our mood. Although chocolate does indeed contain happiness-boosting psychoactive ingredients and antioxidants,[77] any antidepressive effect is more likely simply down to people taking pleasure from chomping on the delectable snack.

The understanding of how our diet affects how good or bad we're feeling and our ability to focus is in its infancy. We do, however, know that what, when and how much we eat and drink and how we view our food is intrinsically linked to our mood, mindset and wellbeing. Even though it's blatantly clear that what we nourish ourselves with affects how we're feeling, the link between nutrition and mental health is often overlooked. When we feel hungry our mood is low, our ability to focus is diminished and we're generally bad company to be around. Similarly, after stuffing ourselves with too many fast-food burgers, cookies and chocolate, the last thing we might want to do is go for a revitalising walk, let alone hit the gym – it's more likely we'll just sink into an inviting armchair. When it comes to the effects of food on our mental health, what might have been less obvious is the profound influence of our long-term nutrition. Diets habitually high in sugar, low in key nutrients and rife with highly processed junk foods adversely impact our wellbeing. If depression or anxiety affect you, making preferable food choices might go a long way to improving your life. Even small dietary changes could provide a motivation and energy pick-me-up enabling you to be more active. And this, in turn, might help boost your mood even more.

Contemplative Food, Nutrition and Mental Wellbeing Key Points

- Avoid processed, high-sugar junk foods.
- Include plenty of fibre-rich foods.
- Eat regularly, but don't overeat.
- Fasting for prolonged periods may be worth considering.
- Include foods rich in omega-3 fats.
- Beware of nutrition claims made by non-expert influencers.
- Hot drinks have feel-good effects.
- Enjoy alcohol in moderation and only occasionally.

Chapter 6

Contemplating the Sustainability of Our Food

It wasn't that long ago when I was paying hardly any attention to environmental issues. Prior to working in the food industry, I was kind of aware that climate change was perhaps something we should be concerned about, but it wasn't a 'me' problem. I'm now slightly embarrassed to admit that I also had little knowledge about how what we eat affects the environment. As a nutritionist I should have been doing a better job. Sadly, however, I've come to realise that, not only should we be trying to reduce the environmental impacts of the food system, but we have an uphill battle as not everyone is on board with the urgency of needing to address climate change and other environmental issues. Indeed, similar to what's happening in nutrition science, misinformation in climate science is rife, too.

It's a frustrating fact that there remains a large number of people who refuse to accept the science: we're facing an existential threat through man-made climate change. Recent data from the Peoples' Climate Vote revealed that a massive one-third of people believe there's no climate emergency.[1] This disappointing statistic has arisen purely out of misinformation and beliefs grounded in assumptions rather than through adherence to a data-based rationale. Yet the research seems conclusive: swarms of scientists from numerous unrelated fields working independently from every corner of the world have demonstrated that the accelerating rate of climate change is man-made.[2] Their results have been replicated and thus provides an example of very strong evidence-based science, more so even than the widely accepted notion that cigarette smoking massively increases the risk of lung cancer. Consequently, the debate should be over, and we've now reached the point that climate change has accelerated to such a degree that there is precious little time left to resolve the situation.

What's also concerning is that many of those who claim to be on board do not realise the scale and urgency. Whether this is the fault of our leaders, scientists, private corporations or educators is up for debate. Despite the cause for alarm, however, there are valid grounds for optimism: the crisis is resolvable as we are not yet at the point of no return. Of course, much of the climate-changing emissions come from power generation, construction, the textile industry, transport and other domains, but the process of feeding humans is a very significant polluter. In its entirety, the food industry contributes around 26 per cent of global greenhouse gas (GHG) emissions,[3] through food production, transport, storage and other processes, as well as those produced from the growing of crops and the farming of animals.

Indeed, concerns surrounding the sustainability of our food system are not limited to its effects on climate change. Issues such as the amount of food waste, land degradation, overfishing and water use are fundamental to any food-related sustainability discussion, as well as how workers and livestock are treated. 'Sustainability' refers to the ability for Earth's biosphere and human civilisation to co-exist. The term encompasses the maintenance of environmental homeostasis (i.e. a stable balance) and involves the conservation of resources to meet the needs of the present without compromising the ability of future generations to meet their needs, with relevance to economic, environmental and social issues. Fortunately, international strategies, such as the 2016 Paris Agreement, national and regional policy, activism, and ESG* actions in private corporations are making some headway.

The intention of a contemplative nutrition strategy is to focus on what we, as individuals, can do in relation to our food choices in order to help minimise our impact on all aspects relating to sustainability. How can our choice of food minimise our impact on land use, overfishing, pollution, food waste and our contribution to the goal of meeting net-zero emissions, while ensuring the optimal health of ourselves as individuals?

First, a Little Bit of Climate Physics

To help make sense of how what we eat impacts on climate change, it might be useful to briefly touch on some of the fundamental physics and introduce some basic climate science terminology.

You might have heard of the greenhouse effect. But what is it? The greenhouse effect is the natural process whereby energy from the sun reaches the Earth's atmosphere, with some of the energy reflected back into space; the rest heats the Earth's surface and is re-radiated. GHGs and water vapour prevent this heat from escaping, and this provides us with warmth. There are several GHGs in the atmosphere with carbon dioxide (CO_2) being by far the most abundant, and nitrous oxide, methane, chlorofluorocarbons (CFCs) and ozone being other main ones. All these gases trap heat, causing the average surface temperature of the Earth to rise, and the greater the quantity of GHGs in the atmosphere, the larger the degree of climate warming. Molecule for molecule, many of the other gases cause more warming

* ESG stands for environmental, social and governance, and refers to actions in respect of ethics and sustainability in businesses; metrics that socially conscious investors use to screen corporations.

than CO_2. For example, methane – which, as you'll soon see, is the principal GHG that we're concerned about when it comes to the food system – can cause up to 120 times more warming the moment it reaches the atmosphere.[4]

When in the atmosphere, these gases stay there for a long time. In the case of CO_2, this could be for as long as 10,000 years. Methane doesn't hang around for anywhere near as long as CO_2: its atmospheric half-life is about a decade. For ease of measurement, we think of all the different GHGs together in units that we call 'carbon dioxide equivalents', abbreviated to 'CO_2e'. This unit takes into account the fact that some gases trap more heat than CO_2 but don't stay around as long. For example, methane is better at trapping heat but has reduced time in the air, so is 21 times more effective than CO_2 in trapping heat in the atmosphere over 100 years.[5] Carbon dioxide equivalents is an imperfect measure as, for example, methane's actual effects can be larger due to the fact that it will likely have more of an impact on raising temperatures in the short term and could have a worse impact than CO_2. It is, however, a convenient and widely used metric.

Up until the mid-1800s, the earth's carbon cycle was roughly in balance, with plant life absorbing as much CO_2 as was emitted. But since we started burning fossil fuels, this balance has been upset and the shift continues to increase, with atmospheric homeostasis continuing to be disrupted and the situation becoming increasingly alarming. The effects of climate change include increased prevalence and intensity of hurricanes and tornadoes and the melting of polar ice. The thawing of permafrost causes a rise in sea levels, and this is leading to the loss of land, which is particularly apparent in poorer parts of the world, i.e. those regions where the negative impacts have greater detrimental effects on populations. Ice melting also leads to an increase in acidification of our oceans and this disrupts the aquatic ecosystem with dire consequences for marine-sourced food.

Furthermore, as ice and snow reflect sunlight, the smaller the amount of the Earth's surface they cover, the less light is reflected and this further contributes to global warming.

Currently, the global annual emissions are roughly 53.85 billion tonnes CO_2e.[6] We need to get back down to net-zero emissions. At net-zero, what we emit in CO_2e is equal to what's removed from the atmosphere by plant life or through technologies. And it's imperative that we get to net-zero. Even if we got down to 'just' 1 billion tonnes CO_2e annually, for instance – despite the fact that this would be a massive success when viewed through the lens of current predictions – the result would still be climate warming, merely delaying the catastrophe. If things continue the way they are, global temperatures will rise by 3.2°C (37.7°F) by the end of this century,[7] a level that experts claim may lead to mass extinctions of large number of species with vast areas of the planet becoming uninhabitable for humans.[8]

And it's not just fossil-fuel-derived emissions that contribute to the increased levels of GHGs in the atmosphere. As permafrost thaws, the carbon that's been stored in the ice for tens of thousands of years is released, exacerbating the problem. Furthermore, as high levels of agriculture leads to more land use, trees and vegetation are being destroyed. Plants require CO_2 for respiration and, in turn, they provide us with oxygen; i.e. they take carbon out of the atmosphere. By destroying these natural carbon sequesters, we're moving things further in the wrong direction. Add to this soil erosion: as soil is eroded it releases stored carbon, and many of the modern arable practices don't allow for it to be taken up again.

Countries all over the world have started to take steps to combat climate change. In 2015, representatives of almost all of the world's nations sat together in Paris and committed to a range of measures. This included targets that would keep the rise in global temperatures to well below 2°C (35.6°F) above pre-industrial levels, as well as agreement to pursue efforts to limit the increase to 1.5°C (34.7°F).[9]

These levels, if successful, would substantially reduce the impacts of anthropogenic climate change. Unfortunately, if current emission levels are anything to go by, virtually no country is on track to meet the 2050 Paris climate change targets. This means that the mean global temperature will most likely increase by at least 1.5°C (34.7°F) in the next 30 years. A 2021 report in *Nature* indicated that the targets set by the Paris Agreement have a drastically low probability of just 5 per cent of being met, and even if planned policies are adopted at a realistic rate, the optimistic probability rises to a mere 26 per cent.[10] In short, much more needs to be done to reduce emissions.

Moreover, the global population, which, at the time of writing, stands at more than 8 billion, is expected to rise to 10 billion by 2057.[11] As the number of mouths that require feeding increases, the demand for the world's natural resources rises exponentially, meaning greater levels of GHGs in the atmosphere and risks in respect of negative climate change outcomes are under further pressure. Globally, numerous strategies and policies are underway that focus on reducing GHGs in the atmosphere, including alternatives to fossil fuels, carbon removal and increasing the efficiency of industry. One key action that would have a huge benefit, and something that each of us can actively influence, is what you choose to eat and how it affects the food system.

What We Eat and Our Food System

The global food system contributes over a quarter of all GHG emissions as a result of land use, crop production, livestock, fisheries and related logistics.[12] Nearly 15 per cent of emissions are from animal-based agriculture via the direct farming of animals and the growing of the crops to feed them.[13] When discussing the effects of our food system, the GHGs produced from all the steps in food production, from farming to post-consumption, should be considered. This includes CO_2 and other GHGs resulting from agriculture, the production of processed foods, packaging, storage and logistics, as well as from food waste. The main GHG produced through agriculture is methane, which is released from the actions of certain microorganisms as they ferment organic material. These microbes are present in wetlands across the globe. Methane is also emitted from the thawing of frozen soils due to warming effects. Another important source of methane is ruminants – cows and other grazing animals – that have methanogenic microbes in their stomachs that help them break down the nutrients from the tough grasses these animals consume. This enteric fermentation results in methane wafting out from both ends of a cow – via their belches and farts – and it's also released in ruminant faeces. Moreover, the manure provides a fermentation site for the methanogenic microbes to produce even more methane. In total, around 32 per cent of anthropogenically derived methane comes from ruminants and manure.[14] Methane is also produced in other areas of agriculture, such as rice paddy fields, which contribute 8 per cent of total man-made methane emissions.[15] It's also released from food waste and human sewage treatment.

Although methane is a much better insulator because it doesn't hang around in the atmosphere for as long as CO_2, tonne for tonne, the resulting measurable climate impact is around 86 times greater than CO_2 over a 20-year period. So, more good news: any decrease in methane will be notably more positive and taking into account the resistance many people have to cutting down on fossil-fuel-derived energy, it makes a lot of sense to include methane reduction as a key target.[16] Indeed, the Climate and Clean Air Coalition – a collaboration of governments and environmental lobby groups – estimates that halving anthropogenic methane emissions over the next 30 years could provide a 0.18°C (32.3°F) benefit in reducing global averages. To put this in perspective, this would mean closing the gap between current temperatures and the 2015 Paris goal by an impressive 45 per cent.[17] This would be a net benefit independent of CO_2. When we consider that over 41 per cent of the methane resulting from human activity is from agriculture and that a large amount of it is also emitted as a result of food production and waste, we can see that a vast majority of methane comes from the global food system.[18] This encouraging news means that something as simple as changing the way you eat will have a very significant impact on limiting the rise of global temperatures.

Without any action, however, things are likely to get worse. Annual global meat production has quadrupled over the past five years, with over 350 million tonnes of meat now produced annually.[19] Add to this, 549 million tonnes of milk and 87 million tonnes of eggs produced each year.[20] Booming demand for animal products inevitably translates into rising levels of domesticated animals that humans use for food. The most recent official figures show that in 2014 the world was home to some 21.4 billion chickens, 1.5 billion cows, 1.2 billion sheep, a billion goats and 986 million pigs.[21] That's a lot of animals raised with the sole purpose of feeding humans and those numbers could increase as the world's population continues to swell, an issue we'll be looking at in the next chapter.

Alarming data released by the Intergovernmental Science-Policy Platform on Biodiversity and Ecosystem Services – a UN-mandated biodiversity watchdog – suggests that human exploitation of land, sea, plants and animals is paving the way for another less-known crisis, which scientists are referring to as the 'biodiversity crisis'. According to a report published in 2019, over one million animal and plant species are at risk of extinction, the largest number in human history.[22] This figure is made all the more sobering by the fact that there are many taxonomic groups for which no firm conclusions can be drawn due to insufficient data being available. Of course, not all of these animals represent domesticated species used for farming but experts warn that almost 10 per cent of animals used for farming or meat production had become extinct in 2016,[23] a figure that's expected to increase. And if you're thinking that having fewer animals on the planet could help to reduce overall emissions from farming, think again. Human societies rely on healthy ecosystems, as do the millions of trees and plants that regularly soak up carbon from the atmosphere to pump out oxygen. In so doing, they mop up as much as 11 billion tonnes of CO_2 each year, equivalent to almost one-third of what anthropogenic activities release each year.[24] Intricate webs link the ecosystem and environmental degradation. The consequence of reducing biodiversity is an exacerbation of the risks posed to us all by climate change.

How on Earth Can We Protect Our Land?

As well as the impact of agriculturally derived GHGs, we've also committed a massive proportion of the planet's land to feed our species. Agriculture utilises over 50 per cent of global habitable land,[25] with the livestock farming system taking up over 83 per cent of this for the animals to graze and for us to grow the crops to feed them.[26] The conversion of natural ecosystems into croplands and pastures on which to grow more food is currently the largest threat to species extinction.[27]

The impact of land-use changes on global GHG emissions is primarily the result of deforestation, intensive livestock rearing, soil degradation and agrochemical application. As more land is used, natural vegetation is destroyed – the very thing that absorbs carbon from the atmosphere. With the rising demand to feed an ever-growing population, and the laws of supply and demand pulling in the same direction, industrial farmers are frantically trying to minimise costs, often using cheap short-term yield-enhancing chemicals and detrimental processes to increase their output. As well as this, the necessity for poorer farmers in developing nations to have a good annual yield has led to significant deterioration of the topsoil of fertile lands.

To avoid a full-scale environmental crisis, scientists and policy-makers are placing their bets on exciting research in regenerative agriculture. Regenerative agriculture is a farming method that aims to restore soil health and to improve the entire ecosystem of a farm by rebuilding organic matter and restoring soil biodiversity. The maths is simple: fertile soils store more carbon. Therefore, increasing the availability of healthy soils means more carbon can be captured from the atmosphere and thrown back into the ground.

Regenerative farming is a crucial weapon in our fight against global warming. It restores the productivity of agricultural lands and secures the livelihoods of farmers. Reversing the degradation of the world's soil is an important way in which we can help slow down climate change while providing the crops we need. Over time, regenerative farming will lead to improved yields, increase resilience of crops to droughts and floods, require fewer agrochemicals, increase farmers' incomes and improve the nutritional quality of what we eat. However, moving over to regenerative farming systems is not without its obstacles and there can be resistance from farmers yet to be convinced of its advantages. In some cases, this reluctance is justified: moving a farm over to regenerative methods often requires a few years until the resulting yields are comparable with those of the previous conventional farming methods on the same land. For many farmers, forgoing their entire salary for even one year isn't a viable option. This is especially true for small-holding farmers in poorer regions who rely on regular yields to feed their families. Policies and charities that could support these farmers while they transition to regenerative agriculture would help; for example, providing sufficiently attractive subsidies that would encourage regenerative farming over more conventional methods.[28] On the other hand, the changeover for wealthier farmers in industrialised nations is far more feasible as they could, for example, convert their farms in stages. The reticence of these farmers may be based on prior habits and beliefs, and their minds may only be changed if it can be demonstrated that, after transitioning, the net result will be a greater income.

Nevertheless, many forward-thinking Western farmers are taking steps to transition. Through an initiative that links regenerative farms and industry players, I had the good fortune to collaborate with one such farm in Northamptonshire, UK. Showsley Farm, run by Carl Krefting and his son Fred, focuses on traditional farming methods involving biodiverse agriculture via a field that they sublet for sheep farming, beekeeping to support pollination, the inclusion of reasonably sized zones that are left unfarmed and adjacent woodland areas. Crops rotated on an annual basis include barley and several varieties of beans. By rotating crops and diversifying plant root structures and organic matter, Carl and Fred are already seeing their soil health being restored. The process also disrupts pest and disease cycles and promotes soil microbes, leading to increased nutrient availability and soil resilience.

As well as there being environmental benefits, switching to regenerative agricultural practices might also lead to more nutritious yields. Modern systems of arable farming favour larger yielding crops with strains selected for their shorter growing time and ease of harvest. Not only do these methods impose harsh environmental consequences, when crops have less time in the soil, levels of key nutrients and phytonutrients can be compromised. Soil health likely has an under-appreciated influence on nutrient density, positively influencing the nutritional profiles of both crops and livestock compared to conventional practices.[29] Traditionally regeneratively grown products have been available only from local independent farm shops, but the larger corporations and supermarkets are catching on and have begun introducing a range of regenerative produce as part of their sustainability initiatives.[30] But while the big guys are playing catch-up, we'd be ill-advised to shun their greens altogether: *all* fruit and veg pack a healthy nutrition punch.

The Limits of Water

As water is technically a renewable resource, it's often overlooked when considering issues relating to sustainability. However, processes like sanitisation and desalination* of water for human use, as well as its use in industry and agriculture, where it's necessary for irrigation and providing hydration for livestock, involves the generation of GHGs. As the world's population increases, with concomitant demands on the amount of water required, the considerable pressures on water resources are further amplified. Furthermore, as global temperatures continue to rise, more water evaporates into the atmosphere despite increasing demand for fresh water in hot regions of the world, which are often areas with higher levels of poverty. To address this challenge, water could be transported directly to these regions, or local governments could invest in bespoke purification plants. Though both of these measures are technically feasible, the problem is both strategies would further aggravate the GHG problem. Moreover, poor management of animal waste pollutes water sources and can lead to algal blooms, eutrophication and the creation of dead zones, meaning less clean water is available. Water should therefore be viewed as a finite commodity.

Annually, each person, on average, uses nearly four million cubic metres of water.[31] That's equivalent to around 1,600 Olympic-sized

* Desalination of water is particularly important in countries like Saudi Arabia where there are few lakes.

swimming pools. Agriculture is the main pressure on renewable water resources. For example, in the spring of 2014, this sector alone used 66 per cent of the total water used in Europe,[32] and food production is responsible for 70 per cent of freshwater use.[33] Animal products typically have a larger water footprint than plant foods when compared both on a weight-for-weight basis and per calorie.[34] Some examples: the average water footprint for beef is twenty times larger per calorie than for cereals and tubers,[35] one litre of milk requires 1,000 litres of water[36] and a 150 g beefburger uses 2,350 litres.[37] It takes 100 times more water to produce a pound of animal protein than a pound of grain-derived protein.[38] So, by consuming a larger proportion of your nutrition from plants, you'll be saving a huge amount of water and, in turn, GHGs.

Watching Our Waste Lines

The tragic irony underpinning the food supply chain is that, while we're needing to feed an ever greater number of mouths, we're throwing a lot of food away. Of all the food produced for human consumption, around 30 per cent is lost or wasted.[39] That's around 1.6 billion tonnes of food each year that fails to meet its objective of providing sustenance for sentient beings. In respect of global anthropogenic methane emissions, landfills and waste management are responsible for roughly 20 per cent.[40] The food we bin, therefore, hikes up levels of GHGs in our atmosphere and, consequently, is a major contributor to climate change.

Indeed, the food that we waste doesn't end with all the rotting veg we throw in the bin. Food can be lost and wasted during harvest, storage, production and transportation. The Food and Agriculture Organization (FAO) of the United Nations defines *food loss* as 'the decrease in the quantity or quality of food resulting from decisions and actions by food suppliers in the chain, excluding retailers, food service providers and consumers'.[41] This includes any food that's discarded or disposed of along the supply chain at any stage prior to the retail level. The FAO defines *food waste* as 'the decrease in the quantity or quality of food resulting from decisions and actions by retailers, food service providers and consumers'.[42] The organisation states that food can be wasted in many ways including fresh produce that deviates from what's considered optimal (i.e. in terms of shape, size and colour), food that's close to, at or beyond its 'best before' or 'use by' date and is discarded and large quantities of edible food unused or left over and discarded from households and eating establishments. The FAO also notes that less food loss and waste would lead to more efficient land use and better water-resource management and have positive impacts on climate change and livelihoods.[43]

In low-income countries, considerable quantities of food are lost before reaching the intended market. This is the result of it spoiling in transit from the hot conditions or from political turmoil causing it to be delayed or diverted and, consequently, thrown away. In affluent nations, around 70 per cent of food wastage happens in the home.[44] For example, the average person in the UK wastes 70 kg of edible food per year.[45] In 21st-century Western societies, most of us have the good fortune to enjoy access to cheap nutrition through being presented with an abundance of food. Consequently, we value our food less. We don't check use-by dates, so food expires and has to be discarded, and if we don't like the flavour of something, we simply lob it in the bin. Such acts would never even enter the mind of someone who struggles to pay for their next meal.

When food is lost and wasted, all the resources required to produce that food – the water, land, fuel, fertilisers, animal life, packaging – are also wasted, and the resulting environmental impact is substantial, contributing as much as 9.3 billion tonnes CO_2e per year to our atmosphere.[46] Chucking one bruised apple in the bin is the equivalent of pouring 125 litres of water down the drain.[47] At current levels, the amount of food that's lost and wasted could feed two billion people.[48]

Of course, there will always be some food that will be lost and wasted through agricultural processes, production and logistics, and much will simply not be consumed for an array of reasons. It's inevitable that there will always be circumstances beyond anyone's control that will result in some food being wasted in manufacture, contaminated or damaged, and much will go out of date. However, we can adopt strategies that will help to minimise these losses. The need for the development of technologies that help to prolong the shelf life of foods while not compromising their quality, improvements in packaging and better logistic and transportation planning is becoming increasingly urgent. Other schemes can focus on incentives. One such initiative involves smart pricing. Wasteless is a company that aims to incentivise at the retail level by dynamically pricing those items with a shorter expiration date at their optimal price point. As food items get closer to their use-by dates, an artificial intelligence technology lowers the item's price increasing the odds of it being sold. This technology allows retailers to reduce waste and increase their profits.[49] Similar, less sophisticated tricks have also been shown to increase sales of potentially perishable food. For instance, carefully choosing the exact location of food within a store can make it more eye-grabbing for customers and less likely that it will go to waste.

Even with the best of intentions, food is routinely lost or wasted at some point along the supply chain; however, the food that does need to be thrown away can be disposed of through best practice, such as using anaerobic digestion facilities that convert food waste into renewable energy. Food producers and retailers can take responsibility for this, and local authorities can offer households suitable refuse solutions that include food caddies whose contents can be collected separately for appropriate disposal. Each of us can be savvier both in the way we shop and how we manage the food in our cupboards. We can all take more responsibility to ensure that the food waste we generate is disposed of correctly.

Plenty More Fish in the Sea?

The way humans are currently fishing is not sustainable and the number of fish in our seas is diminishing. This is because one-third of global fish stocks are overexploited, meaning the rate at which fish can reproduce and replenish their numbers is leading to a significant decline in populations.[50] Scientists have been collecting data on this for years. They use a metric called maximum sustainable yield (MSY), a measurement that refers to the upper limit of the amount of fish that can be harvested without depleting the existing natural resource. Think of this as the ceiling value beyond which fishing stocks will start to decline vertiginously to the point where humans risk having no fish on their dinner plates. Calculations based on MSY have revealed that sustainable stocks have fallen from 90 per cent in the 1970s to 66 per cent in 2017.[51] This is because despite intense fishing activities worldwide, our appetite for fish hasn't budged, with recent estimates suggesting that globally over 200 million tonnes of seafood are produced each year.[52]

To make matters worse, overfishing is not the only reason for the global reduction in fish availability. Global warming and the concomitant acidification of our oceans has sent the marine ecosystem into disarray. A reduction in the pH of the ocean can cause numerous biological effects in many types of aquatic genera. One such effect is the decalcification of shells of shellfish, resulting in their death and depopulation. This effect also manifests in shelled zooplankton, animals that are an early step in the food chain for almost all fish in the ocean, leading to repercussions in population numbers for those fish species that humans consume.

There are, however, a lot of unknowns when it comes to issues relating to sustainability and marine life. In his informative book *The Physics of Climate Change*, renowned physicist Lawrence Krauss makes this point: 'To what extent various larger fish populations, a staple for feeding much of humanity, will survive a combination of increased temperature and increased acidity is really not known. But clearly a collapse of any major fish species would reverberate throughout the planet.'[53]

Aquatic Alternatives

Fish farming is booming. Interestingly, however, it's no modern invention. As far back as 4,000 years ago, Egyptians farmed tilapia in ponds adjacent to the Nile, and the world's first fish-farming manual can be traced back to China about 2,000 years ago.[54] But as the world's population accelerates to 10 billion and the food system is placed under unprecedented strain, fish farming is gaining renewed attention. This is not a coincidence. People like to eat fish and, judging from the staggering amount of fish consumed in modern societies, their appetite for aquatic meat is growing. The average annual rate of global fish consumption between 1961 and 2017 was 3.1 per cent. For perspective, that number is almost twice that of annual world population growth during the same period (1.6 per cent) and is higher than that of any animal food consumed on the planet. It is estimated that in 2018, every person on the planet consumed an average of 20 kg of fish, twice that of half a century earlier.[55] What's driving this trend? While many theories have been proposed, most likely is the flocks of unyielding beef eaters who are now beginning to contemplate scaly alternatives in their diet. In a world agonised by increasing rates of GHGs, this is good news. Studies show that non-terrestrial farming methods have a significantly lower carbon footprint than virtually all grazing meats, especially beef.[56]

Despite its gleaming CO_2 credentials, there's much concern that as the industry grows so does its harmful impact on the planet. An article published in 2021 by the *Financial Times* revealed that aquaculture production ballooned by 527 per cent between 1990 and 2018. During the same period, the amount of captured fish increased by a mere 14 per cent.[57] Despite this vertiginous growth, aquaculture still seems to lag wild fisheries in terms of fish production. According to the FAO, in 2018, aquaculture produced more than 82 million tonnes of aquatic animals while wild fisheries produced 97 million tonnes.[58] But to the delight of fish farmers, aquaculture is projected to continue its relentless growth in the next decade reaching over 100 million tonnes by 2030 before trumping any other fishing method by 2050.[59]

The reason that aquatic animals have an environmental advantage is mostly because fish are cold-blooded creatures that consume less energy to regulate their body temperature than their land-living counterparts. This inherently efficient process allows fish to convert food into body mass more readily than mammals.[60] There is, however, a fundamental problem with the aquaculture production process. In most cases, farmed fish, especially carnivorous species like salmon, are fed feed made from wild fish. A 2022 study showed that this terribly inefficient system wastes a huge amount of nutrients that could otherwise be used to feed humans.[61] A more sustainable solution could involve limiting production of wild-fish-fed species, such as salmon, while increasing farming of species requiring little or no feed, such as mussels and carp. This approach would safeguard local fish supply by diminishing demand for wild-caught fish currently destined to salmon production while sustaining global growth in aquaculture production. As well as this, salmon farms have drawn intense criticism in recent years because they are very energy-intensive and require large amounts of fossil fuels to maintain optimal culture conditions. If the conditions aren't right, the consequences can be catastrophic. In February 2020, a glitch in the filtration system of a large on-land salmon facility in Denmark meant around 227,000 salmon were lost.[62] To make salmon farming facilities more sustainable and cost-effective farmers pack tanks with fish, which maximises their production rate but there are obvious welfare concerns. From a GHG perspective, however, compared to land-based farm production systems, offshore pens offer a potential alternative as recent studies show that their carbon footprint is about half the one of on-land facilities.[63]

Of real interest are molluscs. As demand for seafood continues to grow, this subsector offers some unique environmental advantages. This is because molluscs, such as oysters, mussels and scallops, are not fed nor do their pens require water filtration systems (meaning low electricity requirements). These wondrous creatures actually absorb waste nutrients that are harmful to the ecosystem. Among these is carbon, which they sequester from the atmosphere before locking it up in their shells. Despite these obvious advantages, the production of fed aquaculture fish such as salmon, shrimps, trout and sea bass far outpaces the growth rate of non-fed species.[64] Molluscs don't carry the same demand as fed agriculture species. This might be because people are unaware of their carbon-grabbing advantages. Creating better awareness of the comparative benefit of molluscs could help make aquaculture practices more sustainable as demand for seafood continues to increase. Improving fishing practices and implementing more sustainable fish farming methods could help to reverse the global trend of overfished stocks, preserving biodiversity and, in turn, minimising the environmental impact.

Contemplative Sustainable Nutrition

Considering all of this, how can you eat without contributing to environmental disaster? Let's run the numbers. Meat and dairy account for 83 per cent of the carbon footprint of an average diet[65] but only provide 28 to 30 per cent of the calories.[66] Grains, fruit and veg have the lowest environmental impact per serving while meats from cows, sheep and goats have the highest impact.[67] Vegan and vegetarian diets provide the greatest reduction both in GHG emissions and land use.[68] Even replacing red meat with poultry, fish or pork will reduce the environmental effect of your diet, but to a far lesser degree than moving to plant-based alternatives.[69] When it comes to reducing water use, the data tells us that vegetarian diets are the most effective.[70] Moreover, the impact of plant-based diets leads to more than a three-quarter reduction in the amount of land used.[71] Land that's no longer required for food production could help remove the vast quantities of CO_2 from the atmosphere each year as natural vegetation could be re-established and soil carbon allowed to re-accumulate. As well as this, more land means improved biodiversity and the re-establishment of habitats for endangered species.[72]

Humans don't need to eat anywhere near the amounts of animal-derived nutrition that we typically consume in the West. It won't be long before there are too many people for the current livestock farming system to be able to feed, so any solution will have to involve eating fewer animal products and more plants. Although there is no single definition of a 'plant-rich' or 'plant-based' diet, such diets are considered as those based primarily on foods derived from plants including vegetables, grains, legumes, edible fungi, nuts, seeds, fruit and meat alternatives, with few foods of animal origin. Plant-based diets can also include foods and ingredients created by novel innovations where no animal has been harmed. These include synthetic vitamins and minerals and organoleptic enhancers, such as flavours and sweeteners, as well as new and exciting alternative protein and fat technologies.

As individuals, we have the ability to make a significant impact in the battle against climate change. One of the simplest things we can do is to obtain a greater percentage of our nutrition from plants. Compared to consuming a typical Western diet,[73] switching to one plant-based meal a day will reduce your carbon footprint by 35 per cent; switching to two will cut it by 50 per cent.[74] The effect is a double-whammy: the impact from switching to more plant foods will reduce the CO_2 emissions from what you eat and will also allow carbon to be sequestered back to the land from the atmosphere. Interestingly, however, even among those who are concerned about climate change, many are less willing to change what they eat than they are to undertake other measures of carbon reduction. Data from the Peoples' Climate Vote demonstrated that the promotion of plant-based diets was the least popular of the 18 policies in the survey: only 30 per cent of respondents supported it as an action.[75] What's unclear is whether people think that adopting a plant-based diet doesn't provide enough of an impact against climate change or whether they just aren't motivated enough to eat other stuff. Most of us wish to only help the environment in ways that affect our own lives in the smallest way possible. People are less willing to make relatively benign lifestyle changes – even if those changes may also benefit their health – than they are to support policies like protecting rainforests, renewable energy and climate-friendly farming. This indicates indifference and lack of personal responsibility. We are all guilty of continuing to contribute to the climate-change crisis in some way. Granted, only recently did science reveal the direct link between human activity and climate change, and many people are, no doubt, still unaware of the extent of their impact. But the fact that a large number of people demonstrate some degree of concern about the climate problem, yet are unwilling to take responsibility to do their bit to help mitigate further catastrophe, illustrates an apathetic disregard of their moral culpability.

The collaboration of individuals built the modern world. The same collaboration is what's required to save it. Ultimately, you can mostly be responsible for the choices you make for yourself and your immediate family. Making dietary changes and reducing your own carbon footprint is one of the biggest things you can personally do to help curb the crisis. Collectively, we can make a massive difference and people are influenced by others: by making changes to your own diet, others will see what you do and many may follow. Creating discussion and demand for more environmentally friendly products influences industry and drives favourable competition which, in turn, will help expand choice and reduce prices.

In 2020, some colleagues and I co-authored Huel's *Sustainable Nutrition Report*. In it, we defined 'sustainable nutrition': 'Sustainable nutrition is food that provides essential nutrients in proportions that contribute to good health, is affordable, and produced with minimal impact on the environment to protect biodiversity, ecosystems and communities both for now and for future generations.'[76] This description highlights how sustainable nutrition is integral to a contemplative nutrition strategy. Although, first and foremost, your food choices should consider what's optimum for your own health and wellbeing, our behaviour should also embrace environmental considerations. A key contemplative eating takeaway could be to limit the number of calories derived from animal products to 10 per cent of your energy intake. This strategy, I've estimated, saves around 65 per cent of animal calories when compared to a typical Western diet and reduces a food-derived carbon footprint by 35–45 per cent:* not too shabby. The

---

* This is a reliable estimate calculated through a typical intake of a range of meats, dairy and egg to make up 10 per cent of energy intake,[77] and then compared to the 28–30 per cent Western average.[78]

10 per cent limit is easy: simply reduce the portion size of meat and fish at meals and, on some days, hardly consume any animal products.

Why not try it for yourself? It needn't be overly complicated, and you don't need to walk around with a calculator and food diary or the latest calorie-tracking app. The 10 per cent cap doesn't need to be worked out precisely: simply reduce the amounts of meat and fish you serve yourself, include protein-rich alternatives such as beans, lentils and nuts, stick to having meat, fish and eggs at just one meal in a day, keep your dairy intake to a minimum, and include two or three days a week where you almost entirely stick to plant foods. Of course, you may wish to reduce your intake of animal products even further: maybe go vegetarian or vegan. Contemplative nutrition is fundamentally about making improvements to what you're eating, and by consuming a plant-rich diet with a small amount of animal-derived nutrition you can enjoy lots of variety in your meals.

You may also need to modify your shopping habits. Modern Western culture has this ridiculous notion that it's better to buy too much food than not enough. This mindset encourages wastefulness. Don't overbuy: plan your meals ahead and, if possible, shop locally. This may mean two or three small shopping trips each week, but you'll throw less away and this will also benefit your bank balance. Better still, buy in bulk and freeze. Check use-by dates, both when you're at the store and in your own cupboards. There are numerous ways you can minimise the amount of food you waste. For example, vegetables nearing their use-by date can be combined in multiple ways when used in stews. You'll be surprised how little of an animal you need to chuck in the bin, too. For example, once you've stripped most of the meat from a chicken, use the remnants to make bone broth with vegetables, pulses and potatoes: there are valuable nutrients in the bone marrow. Contemplative sustainable nutrition is about making a concerted effort to reduce the amount of animal-derived foods you consume.

Contemplative Sustainable Nutrition Key Points

- Limit your animal-derived calories to 10 per cent.
- Limit the portion size of the meat and fish you serve.
- Only have meat, fish or eggs at one meal in a day.
- Include two or three days a week where you almost entirely only eat plants.
- Plan ahead: don't overbuy.
- Buy in bulk and utilise a freezer.
- Dispose of food waste appropriately.

Chapter 7

Contemplating the Ethics of What We Eat

When was the last time you looked an animal you eat in the eyes? I don't mean the specific animal you've consumed, but the species. When did you last look a cow in the eyes? A pig? A lamb? Or a chicken? Obviously, this question applies only to meat-eaters, but this thought experiment serves to illustrate the disconnect between modern life and what we eat. In his classic essay *Why Look at Animals?*, English writer John Berger noted the loss of everyday contact – especially eye contact – between ourselves and animals, and that this has left us deeply confused about our relationship to other species.[1] Through looking the animals we eat in the eye, we glimpse the unmistaken familiarity of human traits and emotions such as fear, pain and suffering, or contentedness and courage. On discussing Berger's perspective in *The Omnivore's Dilemma: The Search for a Perfect Meal in a Fast-Food World*, Michael Pollan observed that our ancestors 'built a relationship in which they felt they could both honour and eat animals without looking away. But that accommodation has pretty much broken down; nowadays it seems we either look away or become vegetarians'.[2] The abyss of disconnect has widened over the past several decades and this has contributed to the exploitation of many species by the majority of modern humans. Our ancestors had no such disconnect, nor do hunters, pastoralists and subsistence farmers in the world today.

This chapter explores morality when it comes to deciding what to eat. But what do people mean by 'morality' and 'ethics'? Although you won't be expecting a book on food and nutrition to veer into the kind of controversies that philosophers have been debating for millennia, these concepts are very relevant to any discussion of our food system with the goal of making informed decisions about what to eat. Morality refers to what's 'right' and 'wrong' and this, for a large part, is subjective: what you feel is 'ethically acceptable' might be somewhat different to how I see it. Humans are complex beings – there are many routes to moral judgement and moral action – and we must accept that the lens through which we view the world is unlikely to be the same as that of someone else. This is especially true when it comes to someone from another culture: we must be mindful of our ethnocentric biases, and this especially applies to those of us who have a plentiful supply of food.

Food and Human Rights

Next time you're shopping, consider for a moment that the food's true cost is unlikely to be reflected at the checkout. Recent estimates of the hidden costs of today's food systems range from $12 trillion to $20 trillion annually.[3] For perspective, that's roughly double the total economic value of the global food system.[4] The financial health burdens of poor diets and environmental costs are obvious contributors to these figures, but there are human costs too. Too often, the oppression of other Homo sapiens shuns human rights in favour of economic incentives. Pertinent to any discussion that encourages people to reflect on their food choice is awareness of how those who work in food production are affected.

An overreliance on a food system that's based on monocultures and dominated by a few large companies drives a constant desire to reduce prices and perpetuates a destructive cycle that involves both systemic human rights abuses and environmental destruction. Increasing levels of food production and the associated harms – land and water degradation, biodiversity loss, deforestation and increased greenhouse gas emission – are not only harmful to the environment, but also directly impact humans by restricting access to clean air, water, food, health and livelihoods. Smallholders are losing land to larger companies, communities are being displaced when land grants are being given to large companies. Too often workers' rights are ignored with unregulated health and safety working conditions, forced long hours and child labour. Modern slavery and human trafficking are far more prevalent than many of us choose to acknowledge, with recent estimates suggesting the number of people in forced labour may be as many as 28 million.[5]

With growing alertness of human rights issues and concerns about the environment from consumers, it's encouraging that adopting ESG measures has emerged as a crucial focus for businesses of all sizes to thrive and future-proof themselves. ESG metrics are increasingly influencing companies' growth through driving decisions made at board and investor level. Consumer behaviour and attention to social issues raises the bar and any business not meeting the level of integrity expected by society risks negative perceptions irrespective of legislation; there's nothing a company hates more than negative PR.

Agroecology, too, offers a solution, encouraging a holistic approach to food systems by integrating ecological principles with social equity and offering sustainable practices, with the aim to protect and improve rural livelihoods and social wellbeing. While the responsibility to tackle human rights issues in supply chains lies with the food industry, legislation is imperative, and many of the issues can be simultaneously addressed while tackling climate change.[6] To prevent negative effects, governments could apply a rights-based approach to all food-related laws, regulations, policies and actions. These initiatives could focus on the right to food and the right to a healthy environment, guaranteeing food systems that don't negatively affect the availability, accessibility and quality of local natural resources, or threaten traditional systems of collective use and management of the environment. Governments must ensure that food systems operating in their territories respect the rights of everyone and that they promote substantive equality. Principal issues include labour rights (such as appropriate hours, adequate wages, child exploitation and human trafficking), the right to clean water and sanitation, land conversion and resettlement, indigenous and cultural rights, and food security.

But much of the onus lies with shoppers. Businesses are influenced by customers, most notably by what we buy and boycott, but also by their reputations. We, as consumers, can learn to understand the key issues and educate ourselves. Which foods represent a higher risk and what steps can be taken? For example, some foods are very water intensive, but are produced in water-scarce areas, such as Californian almonds. Knowing what you're eating allows you to choose whether to avoid certain foods or brands. Look for companies that are taking action to understand the issues in their supply chain and addressing them. Those that are demonstrating transparency about their supply chains will have codes of conduct for their suppliers, conduct audits of their suppliers' performance, require suppliers to hold ethical certifications, publish a modern-day slavery statement, among other actions. Through simple shopping choices, you can demonstrate to the food industry, governments and fellow consumers the value you ascribe to fair and just trade. Organisations such as Fairtrade and Rainforest Alliance provide certification and aim to ensure standards in the production and supply of food products and ingredients, addressing workers' rights, working conditions and pay.[7] Favouring certified products demonstrates that you support fairness and equality for farmers amid an era where farming is up against some of its biggest ever challenges – such as climate change, land use and water use – by helping them utilise climate-friendly farming techniques, and supporting gender equality and ethical labour practices. The benefits extend to shoppers in the form of high quality and ethically produced products.

Ethics and Eating Animals

When evaluating whether it's ethical for humans in the 21st century to consume the flesh and produce of nonhuman species, we have to consider why humans eat other animals. The consumption of meat has been a major factor in enabling Homo sapiens to evolve into the successful species that we are today. Our efficient alimentary tracts, metabolisms and big brains are a consequence of the omnivorous diets that we've enjoyed, which have included a significant amount of meat. Interestingly, today, more and more people are choosing to shun meat on the grounds of ethics with particular objections to intensive farming methods. Later we'll discuss if we *should* eat animal products, but first, let's look at why we consume meat.

The Long History of Eating Meat

The consumption of one species by another began hundreds of millions of years ago. The first predators were species of bacteria who preyed upon others by a process called phagocytosis.[8] As soon as these ancient bacteria got hooked on the 'meat' of others, a chain of events led to the emergence of multicellular species, and these continued to consume other multicellular species. Without bacterial phagocytosis, there would be no animals, plants or life as we know it today. The first true carnivores appeared in the oceans of the late

Precambrian era. Although we're unsure what this predator was, fossils of its prey dating to around 550 million years ago have been recovered.[9] Ocean carnivores developed into land carnivores, which became the dinosaurs – both carnivorous and herbivorous. After the dinosaurs had been wiped out, the earliest known primate, Purgatorius, appeared, which was herbivorous.[10] The vegan diet of primates continued for tens of millions of years, save the occasional accidental eating of insects and worms. When, around seven million years ago, one of the primate lineages split into two,[11] one becoming today's chimps and bonobos – who are mostly plant-eaters, but regularly enjoy insects and occasionally small mammals – and the other leading to our early hominin ancestors. Although in our lineage the first species consumed some animals, it wasn't until around 2.3 million years ago that meat-eating became a notable contributor to daily calories in early humans.[12] Through increased group collaboration and the use of stone tools, we became more choosy and could hunt – and share – larger game like wildebeest and zebras. Meat was no longer a rare delicacy and regularly appeared on the menu. Another crucial change occurred around 400,000 years ago when humans started to manipulate fire to cook meat.[13] This caused a massive evolutionary transition where the digestive process became more efficient, allowing precious nutrients to be used for cognitive development. Through animal agriculture, from around 12,000 years ago right up until the 21st century for many human populations, the prevalence of meat on the menu has increased.

Consequently, the consumption of meat and other animal products has been paramount to human development, allowing us to experience the lifestyles we enjoy today. However, evolution has shown us that our ancestors didn't always rely on meat-eating for sustenance and humans have the ability to acquire sustenance from a huge range of foods. So, the trillion-dollar question is, do we still *need* to eat meat today?

Meat-Eating Today

We kill a lot of animals to feed ourselves. In 2022, an estimated 75 billion chickens, 3 billion ducks, 1.5 billion pigs, 515 million turkeys, 637 million sheep, 504 million goats and 308 million cattle were killed for meat production.[14] These statistics, however, give no indication as to how the creatures were farmed and slaughtered. Concerningly, some pundits predict that total meat consumption will rise by 76 per cent by 2050,[15] which includes a doubling of poultry consumption, a 69 per cent rise in beef-eating and 42 per cent more pork products being consumed. This forecasted increase raises concerns about how we're going to mitigate climate-change concerns. For a minority of populations, there's no alternative to diets very high in meat and other animal products. Nomadic pastoralists in arid environments and traditional Inuit communities in the Arctic rely on farming or hunting animals. Conversely, people in some regions may be too poor to buy more than small amounts of meat. For a large proportion of the global population, the price of meat today, relative to their average income, is the lowest it's been in history.[16]

Our decision to consume meat is influenced by many factors, most notably price, availability, convenience and its social and cultural value, as well as economics and politics. Our innate food preferences have evolved out of environments where food was scarce and life was hard, and this has led to our diets being shaped by biological impulses interacting with psychological cues: we're wired to enjoy chomping on animal flesh. Although our intrinsic desire for protein-rich meat once promoted survival, today the amount of flesh we consume contributes both to ill health and environmental destruction.[17] The food we acquire and share is influenced by our beliefs and values, and is part of how we've constructed our identities. In addition to deliberative thinking, the nonconscious mind, influenced by the force of habit and societal norms, influences how we consume meat.[18] We try to justify our consumption of meat to ourselves. Studies have shown that people believe that it is 'natural, normal, necessary or nice'.[19] But it's precisely because most of us view meat consumption as a 'normal' part of our diets – often the routine centre of the main meal – that the *choice* to consume it goes largely unexamined. Yet, social norms can and do change, and this process can be aided by the coordinated efforts of society, health organisations and governments, as has been the case with attitudes towards, say, cigarette smoking compared with as recently as 30 years ago. Despite this, the extent to which the state should factor in considerations like animal welfare when reviewing legislation concerning the production and sale of certain types of meat continues to be controversial. There are technical and philosophical challenges in assessing an animal's quality of life, complicated by the fact that modern livestock breeds are the result of many generations of selection over thousands of years for economically important traits. Moreover, intervention by governments to protect certain wild animals from being hunted for food is often a contentious issue: such as where consumption of 'bushmeat' is culturally ingrained. Most countries have legislation that prohibits the production and sale of certain types of meat, based on animal rights or conservation considerations, but there's a lack of consensus as to how strict legal interventions should be.[20]

Down on the Modern Factory Farm

Cows grazing amid a vista of green pasture, chickens merrily roaming around a farmyard or snuggling happily in a straw-lined wooden coop, lambs frolicking around their mothers and contented farmers carefully tending their livestock. These idyllic scenes invoke feelings of the beauty of rural simplicity and all the bliss associated with living with nature. Childhood farm pictures are equally jovial, with cartoon sketches of farmyard animals smiling gleefully, enjoying their happy lives. Intuitively, we connect these feelings of agrarian purity with the meat, dairy products and eggs we enjoy. The reality, however, is stark and far removed from this notion. The majority of livestock animals are reared very differently. Modern industrial farms are factories with animals viewed as equipment imperfectly designed to convert fuel into profitable commodities. To quote the Australian philosopher Peter Singer: 'The dairy cow, once seen peacefully, even idyllically, roaming the hills, is now a carefully monitored, fine-tuned milk machine. The bucolic picture of the dairy cow playing with her calf in the pasture is no part of commercial milk production'.[21]

In his seminal book *Animal Liberation*, Singer depicts the conditions of modern factory farming through a factual and evidence-based lens. Detailing the practices in an objective way allows us to take a rational perspective when looking to make better food choices. However, despite Singer's attempt at a non-emotive narrative, his description leaves readers shocked at the extent of the gruesome treatment to which animals are subjected. Although some people might be oblivious to the fact that their meat and milk originate from animals raised in such abhorrent circumstances, many others exhibit a degree of cognitive dissonance despite an awareness that animals are reared in such conditions, choosing to pay little attention to the issue. Of course, those who work on factory farms or elsewhere in the supply chain shouldn't be viewed as 'bad' people. The majority are well-meaning, and more likely fail to make a connection to the harsh reality of what they're doing. But we should be talking about modern livestock farming; a position asserted by the historian Yuval Noah Harari in his foreword to the 2015 edition of *Animal Liberation*: 'given the immense power humankind wields over all other animals, it is our ethical responsibility to debate it carefully'.[22]

Modern domesticated animals are the result of thousands of years of selective breeding that favour characteristics for the optimal production of meat, milk, eggs, leather and animal muscle-power. Yet these same animals have retained the inherited physical, emotional and social needs of their free-roaming ancestors. These traits, now redundant on modern farms, are ignored by farmers who lock them in tiny cages, mutilate their beaks, horns and tails, separate mothers from their offspring and feed them diets for which they are ill-adapted. This flies in the face of the fundamental principles of Darwinian evolution, which otherwise maintains that all instincts, drives and emotions have evolved in the interest of survival and reproduction. A farmer takes a young calf, separates her from her mother, puts her in a tiny cage, vaccinates her, provides more than sufficient food and water, and, when she's old enough, artificially inseminates her with bull semen. From a Darwinian perspective, the calf's 'needs' are completely taken care of, but her subjective reality is that she still has impelling urges to bond with her mother and play with other calves, let alone the freedom to roam and stretch her limbs.[23]

Ten thousand years ago the chicken was a rare bird confined to small niches of South Asia. Today's domesticated chickens are the most widespread birds on the planet with billions residing in farms on every continent bar Antarctica.[24] Most of today's chickens, however, aren't clucking merrily around a peaceful farmyard. A modern industrial egg farm contains thousands of birds reared several to a mesh cage.[25] Under these conditions, deprived of their freedom, birds are so stressed that they have their beaks clipped to prevent them from pecking their cellmates.[26] Many suffer broken legs resulting from the cramped conditions, others experience sores from the toxic ammonia-rich excrement.[27] They reside in these highly distressing conditions until their egg-laying days are over, at which point they're killed and their carcasses used for products like chicken nuggets, pies, commercial soups and pet food.[28] Other breeds are reared for their meat. Known as broilers, these chickens are fed on a diet of primarily corn with additional protein, essential fats, vitamins and minerals,[29] and are inhibited from exercising. The goal is simple: to maximise growth in the shortest period possible. Typically, broilers are slaughtered at around seven weeks, at which time they've grown from a mere few ounces in weight at birth to four or five pounds.[30]

Pigs are animals with a high level of intelligence, possibly more so than dogs (it's possible to rear pigs as human companions).[31] Their habitat, therefore, requires special considerations in order for them to be stimulated, and naturally pigs are social creatures who roam and forage for food. On factory farms, however, a pig's life is bleak with nothing to do but eat, sleep, stand up and lie down. Confined to pens, bored, unhappy and highly stressed, they gain weight. Pregnant sows are confined to stalls where they're unable to turn or move more than a few inches back and forth, and on giving birth, they're immobilised until their piglets are weaned and taken away to lead their own short, miserable lives.[32]

Some intensive cattle farms permit cows to live outdoors with their companions. However, they can be confined to enclosures, and the conditions in which they're reared are similarly distressing. Calves, removed from their mothers shortly after being born, are fed diets of corn, soy, fish remains, micronutrients, hormones and antibiotics to promote rapid growth cheaply and in the shortest time possible. Such diets are not ones that their digestive systems have evolved to tolerate and, consequently, it's not uncommon for them to experience gastrointestinal distress. On intensive dairy farms, heifers are reared on special formulae containing hormones designed to maximise milk production.[33]

Of all the forms of intensive animal farming, few are as morally repugnant as veal production. Veal is a calf meat delicacy favoured by elite restaurants and their patrons due to its tender nature and pale colour. Originally, calves were taken from their mothers immediately prior to being weaned off their mothers. A calf that's yet to consume grass gives rise to a paler and tenderer meat. The problem was that calves at just a few weeks old are small and, consequently, don't produce much veal. Considering that farmers sell veal for an inflated price, it makes sense to maximise the yield. This has led to innovative veal farmers coming up with a solution: by taking calves away from their natural conditions, they can be kept alive for longer – up to four months – without their flesh becoming red. Modern veal is produced from calves who, just days after being born, are taken from their mother and put into a tiny pen where their movement is severely restricted and they're unable to turn around. The narrow stalls restrain the infant so much that he's unable to exercise at all, so his muscles are kept weak and the meat tender. Stalls are devoid of bedding because consuming straw darkens his flesh. The calf is fed a liquid formula based on low-fat milk powder with added vitamins and minerals. He's permitted only a sufficient amount of iron to allow him to survive until it's time for him to be slaughtered, while remaining anaemic to ensure pale-coloured meat. By keeping him in warm conditions and deprived of water, he's encouraged to drink more of the liquid feed to increase his weight. The more flesh he grows, the greater his market value will be.[34] The farming of calves for veal is as abhorrent as it is unnecessary: there's no benefit from consuming veal meat over beef other than snob-value since it's more expensive. In fact, due to its lack of iron, veal is nutritionally inferior.

Is Factory Farming Ethical?

Despite the gruesomeness, we should acknowledge that there might be arguments – however questionable – that favour such practices. Some people feel that it's unclear if animals do suffer, at least in a similar way to humans. Advocates of factory farming feel that animals are in servitude to humans who, they claim, sit at the top of the food chain. It's a human right to use animals in any way desired for the provision of sustenance or even for enjoyment. The welfare of nonhuman species need not factor into any considerations, unless an issue is of benefit to the quality or profitability of the commodities they produce. Moreover, they argue, how else are we going to provide sustenance to several billion – and counting – humans? Do we not need efficient methods of production to provide affordable protein for as many people as possible? These claims are highly contentious.

The animal behaviourist Temple Grandin, while not wholly against intensive commercial farming, feels that many of the current practices are morally questionable: 'Using animals for food is an ethical thing to do, but we've got to do it right. We've got to give those animals a decent life and we've got to give them a painless death.' She points out that the majority of animals we eat don't suffer a humane death in nature; most are torn apart by carnivores, rarely dying of old age. 'Nature is cruel but we don't have to be. I wouldn't want to have my guts ripped out by a lion. I'd much rather die in a slaughterhouse if it were done right'.[35] The obvious difference here is, of course, in nature animals often lead relatively long lives or, at least, they have the opportunity to. In factory farms, they're typically killed after a few weeks.

The claim that we need factory farming in order to feed a growing global human population is easily disputed. As we saw in the last chapter, when it comes to calories produced and the amount of land used, livestock farming is massively inefficient. For example, it takes nine crop calories to produce one chicken calorie, and despite this appalling inefficiency, of all livestock, chicken is the *most* efficient.[36] What many fail to realise is that factory farming makes less sense economically, too. A calf, for instance, grazing on a rough pasture that's unable to be used for growing crops, enables something from which humans are unable to extract nutrients – grass – to be converted into human fuel. The result is a net gain of protein and calories for humans. The same calf in a feedlot has to be fed crops such as corn and soy, produce that could otherwise be fed to humans. Industrial livestock farming does little to help feed people in under-developed regions; it serves merely to satisfy the meat-craving impulses of already well-nourished people, to make their animal produce more affordable and to increase the profits of the corporations running the farms. The fundamental existence of factory farms is because humans get a whole load of pleasure from eating meat. The satisfaction gleaned from munching on a cheap Big Mac far outweighs any concern we have for the suffering of the cow that provided the patty. The harsh reality is, every time you purchase a product that involves an intensively farmed animal, you are condoning such conditions.

There are few who outright defend the factory farming of animals and the majority of people are content to continue to consume the produce from such environments. The way we buy our animal products distances the consumer from the harsh reality of the trauma and suffering factory farms commit to nonhuman species. We're sold meat in plastic packages, milk in cartons, yoghurt in tubs, butter in foil wrappers and eggs in boxes. Language further separates the connection to living, breathing creatures: 'beef', 'veal', 'pork' and 'venison', for example, make no reference to the species we're eating. Interestingly, 'lamb', 'chicken' and 'duck' seem to have escaped the lexical dissonance. In our minds, when we eat meat, even though we know where it comes from, we fail to join the dots between the slaughtering of sentient beings and the plated pleasure in front of us, let alone the harsh lives endured by the animals while being reared. Many people might be aware of the appalling practices and they may feel a little uncomfortable about them, yet they continue to bolt down their burgers and chomp on chicken wings. While it's likely that most people don't actively condone these hideous practices, they simply don't care enough to change their own behaviour, and many fail to acknowledge that the hedonic sensation they experience from a single meal is worth more to them than the suffering of a sentient being. This brings us to a key question when it comes to the ethics of eating meat and other animal products: do animals suffer?

Pain and Suffering in Nonhuman Animals

In *Animal Liberation*, Peter Singer refers to 'speciesism',[37] a term coined by British psychologist Richard D. Ryder in a 1970 pamphlet about animal experimentation.[38] Ryder wrote: 'Since Darwin, scientists have agreed that there is no "magical" essential difference between humans and other animals, biologically-speaking. Why then do we make an almost total distinction morally? If all organisms are on one physical continuum, then we should also be on the same moral continuum.' Drawing a parallel with racism – the discrimination of humans based on their skin colour or where they were born – speciesism refers to the prejudice against animals from a different species classification. Singer also cites the 18th-century philosopher and founder of utilitarianism Jeremy Bentham who, when referring to the rights of animals, wrote 'The question is not, Can they *reason*? Nor Can they *talk*? but, Can they *suffer*?'[39]

To most people the idea that animals can feel pain seems rather obvious. The short, sharp cry of a dog when stepped on seems to universally silence even the loudest sceptics of animal pain theories. And yet controversy exists as to whether animals, especially farm animals, feel pain. Perhaps confusion in this area stems from the fact that, for most of their history, humans have shown a stubborn indifference, not only towards animal pain but even towards their own pain. Until at least the 19th century, pain was valued positively and considered an exemplary means of punishment, particularly in the military setting where accepting pain was often promoted as a sign of masculinity.[40] Advances in our understanding of the physiology of pain made it possible to radically transform our perception of it and meet society's increasing demand for treatments to attenuate bodily suffering.

The research on animal pain is fragmentary and hard to fetch, particularly in the context of farm animals. This may well be a consequence of the bullyish behaviour of the meat industry, for which there's likely little to worry about when it comes to animal suffering. Nevertheless, there is growing consensus that animals do feel pain, not least because pain is an evolutionarily advantageous physiological mechanism that alerts living beings that something is wrong. From a Darwinian perspective, it is easy to see the survival value in feeling pain. In the context of farm animals, strong evidence demonstrating animal suffering comes from physiological studies that measured the activity of the hypothalamic–pituitary–adrenal (HPA) axis in relation to pain. The HPA axis is activated in response to internal or external stressors, stimulating the adrenal glands to produce cortisol. In pigs, farming methods, such as tail docking, teeth resection, iron administration and castration, have been demonstrated to increase cortisol levels.[41] Worse still, the type of pain that some farm animals experience is not temporary: studies have shown that debeaking in chickens causes cardiovascular and behavioural changes that can last from weeks to months,[42] and zero-grazing confinement systems increase the rate of long-term diseases in cows compared to free-ranging animals.[43]

Our brains have been programmed so that the sensation of pain tells us *what you've just done is bad; don't do it again*. From this perspective, it's not clear that intelligent species, like humans, are more capable of suffering than unintelligent ones. Indeed, the reverse might be more plausible. An unintelligent species would require *more* pain in order to ram home the point not to do that 'bad' thing again. Whereas an intelligent species might be satisfied with a 'red flag' in the brain, one with little capacity for reason might need a harder prod. It follows that less intelligent creatures might experience a greater degree of suffering than humans. There might, therefore, be a negative correlation between brain capacity and pain and suffering; a perspective that would favour the better treatment of animals. Moreover, not only do animals feel pain and, consequently, suffer from the physical discomfort associated with injuries inflicted through factory-farming practices, they also experience psychological suffering from the absence of parent–child bonding, being isolated from their social group and being restricted from roaming.

The existence of pertinent legislation to safeguard animals from unnecessary pain further strengthens the case for the universal recognition of animal suffering. In the UK for instance, the Animal Welfare Act 2006 protects farm animal welfare by preventing unnecessary suffering.[44] In fact, awareness of animal suffering has reached such heights in our society that scientists have started to toil with the idea of tweaking neuronal pathways to numb pain in farm animals. The wacky idea behind this is that because animals do feel pain, silencing relevant pathways in them would minimise pain, thus making farming practices a bit more acceptable. A 2010 *New York Times* article suggested that developments in genetic engineering could shield humans from that annoying sense of guilt that soaks our meat-heavy menus. Because mice can now be engineered without pain receptors, the theory goes, it might one day be possible to consume genetically engineered pigs or cows that lack genes to feel pain.[45] Although the idea that genetic tools may one day knock out pain and discomfort in farm animals (and humans) seems far-fetched, convincing evidence shows that reducing pain in livestock could have considerable benefits for society. A 2005 study showed that the annual total cost of foot-rot in sheep – a relatively common disease characterised by skin inflammation between the toes – was estimated to cost the UK £24 million.[46] Prompt treatment of this condition has been shown to ensure better productivity of the ewes, meaning more were lambed and profits increased as a result.[47] These figures tell us that by reducing pain in animals, governments and farmers can eschew large financial losses, increase productivity and reduce the use of medicines, such as antibiotics. These are some of the reasons why animal welfare scientists and advocates are increasingly calling for a shift in farming methods to implement a more equitable and pain-free food system. Yet, while removing pain might seem – to some, at least – a step in the right direction towards improved animal welfare, numbing a creature's pain in no way tells us that it will lead a good life or feel healthy; anyone who's been medicated with prescription painkillers might attest to this. We have no way of knowing the subjective feelings of the animal, which calls into question the ethics of such practices.

Overwhelming evidence has shown us that cows, pigs, chickens and other animals reared on modern factory farms not only feel pain but also have a subjective experience of suffering. This leads us to consider the philosophical question: *would it have been preferable for the animals to have never existed?* Without factory farms and the practices that create their livestock – artificial insemination, cramped conditions, overfeeding – it would mean that billions of future creatures will have escaped a miserable existence.

The Welfare of Fish

We've looked at both the benefits of oily-fish-derived omega-3s on our health and the sustainability issues surrounding marine aquaculture. We've seen that while some seafood sources have a lower environmental impact, others contribute to overfishing. Here the focus is on fish farming and its ethical considerations. Similar to terrestrial factory farms, aquaculture systems have been designed with economics at the forefront. Consequently, there are concerns about the cramped conditions of aquaculture farms. In many cases, fish are reared in what some consider to be low-welfare, underwater factories. With so many salmon crammed into tanks and cages, while each around 75 cm long they are sometimes allocated space equivalent to just a bathtub of water each, severely restricting their natural behaviours.[48] As migratory creatures who naturally swim great distances at sea,[49] salmon are forced to swim in circles around the cage, rubbing against the mesh and each other. Only relatively recently has it been realised that most fish species have developed senses and lead social lives, communicate and cooperate in groups, and even exhibit cultural traits.[50] Although fish brains are organised very differently from mammals, they may lead complex emotional lives and studies have demonstrated that fish express fear, display moods and feel pleasure.[51] Even though the subjective experience of fish is probably very different to that of mammals, that doesn't mean that fish cannot suffer in their own way. Yet, not all researchers are convinced that fish actually feel the sensation of pain. Brian Key, an Australian professor of biomedical science, claims that fish don't have the brains for pain, and feels that in experiments that supposedly demonstrate fish's capacity for pain, fish are just automatically reacting to harmful stimuli without their brains registering it as pain the way ours do.[52]

There are also concerns over routine slaughter methods. Fish are removed from water and left to die in distress from asphyxiation, often fighting for their lives for up to an hour. Other methods of slaughter include submersion in a mixture of ice and water, exposure to carbon dioxide and being gutted alive.[53] All these methods likely lead to considerable pain and fear, indicated by fish's violent struggles, often suffering for prolonged periods while remaining conscious and aware of pain.

Carnivorous farmed fish are fed feed made from wild fish. It can take up to 350 wild-caught fish to raise a single farmed salmon, meaning that fish farming actually increases the pressure on wild stocks.[54] One report found that about one-fifth of all wild-caught fish are turned into fishmeal and fish oil to feed farmed fish.[55] These wild fish also have welfare considerations. The rationale put forward for fish farming is simple: it's a solution to the problem of overfishing. But this is only true for some of the species and it's certainly not true for the wild fish that are caught to feed the farmed carnivores. These days, over half the fish that people consume comes from aquaculture – mostly industrial, underwater factory farms. In 2015, in Europe alone, approximately one billion fish were produced in underwater cages.[56] Currently, it's virtually impossible to find 'higher-welfare' wild-caught fish, and sustainability certification schemes, such as MSC,*[57] don't cover animal welfare.[58] Unless explicitly stated otherwise, some groups claim that it's reasonable to assume that most of the fish sold in supermarkets, restaurants and other outlets have suffered at some point during rearing, capture or slaughter.

Contemplating Ethical Eating

This is all just a mere snapshot of some of the key issues concerning the ethics and farming of nonhuman animals for human food. Nevertheless, what's been revealed is that, when it comes to the factory farming of animals, we must do whatever we can to move away from these practices. We should be taking action to make these facilities unviable so that they can be closed down and have their doors bolted. But even if you're still unconvinced that modern factory farms are unethical and you remain sceptical that intensive animal farming has a notable impact on climate change, another disturbing issue will be guaranteed to at least turn the head of even the most devoted burger binger and fried-chicken fan.

---

* The Marine Stewardship Council (MSC) is an international non-profit organisation whose mission is to stop overfishing.

Pathogenic Farming

During a holiday in Cambodia, my wife and I had the pleasure of visiting a local indoor food market: the perfect opportunity for a nutritionist to learn about the local food shopping habits. The market area was not dissimilar in size to a UK supermarket, but without fridges, freezers and air conditioning. In fact, although situated in the shade, the market was very warm. On offer was a huge range of fresh produce – plant and animal; dead and alive – and little in the way of processed food. What troubled me, however, was that the different animal products were being sold in close proximity to each other. While one vendor sliced up a carcass to sell as steaks, his neighbours displayed live fish, crustaceans, arachnids and insects, killed and bagged to order. It immediately struck me that an environment like this would be an ideal breeding ground for pathogens and infection.

The risk of zoological pathogenic spread to humans is no less of a concern in modern factory farms with animals reared in incredibly confined proximity and steeped in excrement. Indeed, disease risk is so significant that antibiotics have to be routinely administered to prevent the loss of livestock and the associated financial cost. In the US, around 80 per cent of antibiotics are produced for animal use, massively exacerbating the problem of antibiotic resistance.[59] How long might it be before antibiotics cease to be effective against dangerous human pathogens?

The risk of disease also stems from what livestock are fed. Following the slaughtering of livestock, the parts of the carcass that are unable to be sold as meat for human or pet consumption, may be fed back to the livestock.[60] Creutzfeldt-Jakob disease (CJD), a rare and fatal disease leading to brain degeneration, is caused by an abnormal infectious protein, known as a prion, which accumulates at high levels in the brain causing irreversible nerve damage. Prions can be highly infectious but, unlike bacteria and viruses, are untreatable.

CJD occurs when the infectious agent jumps from an infected bovine host – where it's more familiarly known as 'mad cow disease'* – to a human. The prion is transmitted either through eating infected beef or exposure to workers. At the peak of the UK mad cow disease epidemic in the early 1990s, an estimated 180,000 cattle were affected, and to prevent the disease spreading, more than 4.4 million cattle were slaughtered.[61] The way livestock are reared and forced into cannibalism hugely increases the risk of them, and consequently humans, becoming infected. Thankfully, to date, incidence of prion disease has been rare, but there's a scary realisation: should a future prion agent be released, the resulting effects could be catastrophic, making COVID-19, Spanish flu and HIV pale into insignificance.

Ethics and Flexitarian Eating

Recognising that the systems of modern factory farming risk disease, contribute to antibiotic resistance, have environmental consequences and impose suffering on animals tells us that we should be taking all necessary steps to move away from these practices. For many, ridding factory farming from the global food system is a moral imperative. While there's still a long way to go and billions of animals will continue to suffer, there's hope on the horizon. Policymakers in some countries are signalling a desire to improve farming methods. The UK, for example, recently published a policy paper that seeks to support farmers in transitioning to higher-welfare practices.[62] Any attempt to change behaviour can be challenging and there are justifiable concerns about the effectiveness and ethics of trying to manipulate populations. This is especially true when it comes to the consumption of meat, which is something that's both pleasurable

---

* The term 'mad cow disease' has been coined because the cow's brain has been affected. The official name is bovine spongiform encephalopathy (BSE).

and, as demonstrated, value-based. Libertarian paternalism involves the design of policies that aim to nudge individuals towards better choices without limiting their liberty,[63] but these can only be enacted in conjunction with honest education and is not something that will reap results overnight. Realistically, it will be decades before we see the end of factory farming, with its demise dependent on alternatives to meat. In the meantime, anyone objecting to these practices who wishes to live their life in a way that's honest to their values should try to avoid eating animal products produced by big corporations. By continuing to buy and consume produce from factory farms, you're supporting them.

Encouragingly, many people are choosing to cut down on or to exclude animal products altogether, and vegetarian diets are more popular than at any time in the past. Some people who identify as vegan follow completely plant-based diets, whereas others also avoid wearing wool or leather, shunning any product involving the exploitation of animals. In the two-year period from 2019, the number of Americans following a vegan diet rose from 0.4 to 3.5 per cent, and the number of Brits has quadrupled over the last five years to around 1.2 per cent of the UK population. Globally, the total number of vegetarians and vegans as of 2023 was as many as 22 per cent, according to a 2024 survey.[64] A 2020 report published in *Nature* noted the varied prevalence of vegetarianism with recent polls indicating around 5 per cent of Americans, 8 per cent of Canadians and 4.3 per cent of Germans follow vegetarian diets. India has the most vegetarians, with around 30 per cent of the population. Veganism is less common: its prevalence is around 2 per cent in the US and under 1 per cent in Germany.[65]

Crucially, you don't have to be a vegan, or even a vegetarian, to make a difference. You need merely to moderate your consumption of meat, milk and eggs, especially those produced by modern industrial farming methods. Flexitarianism is a style of eating that encourages mostly plant foods allowing the limited and occasional intake of animal produce. Many vegans, however, don't feel flexitarianism is ethically sound, seeing adherents as not being honestly committed to the welfare of nonhuman animals. This criticism is unfair and assumes that a flexitarian's motivation should be solely based on the immorality of eating animals, whereas someone's desire to reduce the amount of meat they consume may be based on health or environmental concerns or as a protest to factory farming. Many flexitarians are former vegans who felt their own wellbeing was suboptimal when they adhered to a 100 per cent plant-based diet, possibly as a result of a low iron status. Haem iron, a form of dietary iron primarily found in meat- and animal-derived products like eggs, is more bioavailable than non-haem iron, the type mostly found in plants.[66] Some people, women in particular, risk iron-deficiency anaemia, a condition characterised by weakness and lethargy, which can be easily improved by including haem iron obtained through including a small amount of animal foods. (Although a vegan diet can provide sufficient iron, even for those most at risk of iron deficiency, a fully plant-based diet will need careful planning and should also include foods rich in vitamin C (or as a supplement) as it promotes the absorption of iron.)[67] For others, moving to a completely plant-based diet can be too much of a deprivation of the pleasure derived from eating meat. Flexitarianism provides a stepping-stone for those intent on eating fewer animal products and supports the quest to reduce the amount of meat in the food system by increasing the demand for nutrient-rich, enjoyable plant-based options.

When it comes to animal produce, what and how much should we eat? Here, opinions vary significantly. At the extreme, a vegan diet has the greatest impact on reducing carbon emissions, land and water use, and mitigates the pitfalls of having to navigate ethical concerns. But veganism isn't for everyone, and you can make a huge difference simply by limiting your consumption of animal products to 10 per cent, as demonstrated in the last chapter. It's also reasonable to commit to avoiding the produce of factory farms and to consuming only eggs from free-roaming chickens,* meat from hunted game or traditional farms where animals have been allowed to roam relatively freely and, at a minimum, have led lives absent of human-induced suffering. Granted, such produce can be more expensive, but the cost is more than offset by consuming less. Dairy plays a large part in Western diets, and milk and milk constituents – such as whey, casein and lactose – make frequent appearances on food labels. Consequently, excluding dairy entirely from your diet could be quite a challenge. If you are concerned about the welfare of dairy cattle, a practical strategy might include reducing the amount of cow's milk, yoghurt and cheese you purchase, sticking only to butter from grass-fed cows, and buying milk labelled 'organic' whenever possible. Although organic produce from a nutritional and quality perspective has little merit, there may be benefits when it comes to organic dairying. Cows are housed in groups and are allowed to graze in pastures during warmer months, with access to a more natural, forage-based diet. Although they produce less milk, they experience lower levels of metabolic stress and antibiotics are seldom required.[68] There are still welfare concerns, however. Cattle may still be mistreated and, in any form of dairy farming, calves are often removed from their mothers early in their lives. Fortunately, these days, an abundance of non-dairy alternatives are available. Plant-based milks, such as oat, soy, almond and coconut, along with soy-based alternatives to yoghurts, are widely available.

---

* The term 'free-range' eggs applies only to non-caged chickens who are permitted access to the outside. Often 'free-range' birds are crammed hundreds to a barn with only a small access to the outside. Truly free-range chickens should be permitted to freely roam in smaller groups in large outdoor areas with access to shelter.

What about seafood? Here, the decision is less clear. It's not easy to firmly establish the extent of fish suffering, and the environmental impacts of fishing and fish farming are inconclusive and certainly hazier than aquaculturists would have us believe. We've seen that having molluscs on the menu might have some merit, but what about insects? Throughout history, and even today, numerous cultures have enjoyed grubs for their grub, and globally more than two billion people today regularly munch on insects.[69] Even though more than a thousand species are eaten around the world, they hardly feature in the diets of those in affluent nations. While many modern Westerners might turn up their noses at the thought of eating crickets, nutritious protein-, fat- and micronutrient-rich bug-derived products are appearing on supermarket shelves.[70] Bugs for consumption are typically bred in large-scale factories and the global edible insect market is predicted to exceed $17.6 billion by 2032.[71] The key consideration here, however, is not if we're put off by the thought of gobbling insects, but how people feel about the ethics of eating these creatures.

When it comes to deliberating the ethics of what we eat, it's not so black and white. The Good Food Institute (GFI) is an international nonprofit that aims to find novel methods of alternative proteins and to make them the default choice. The GFI doesn't actually focus on the ethical debate of eating animals; rather, it considers the contribution of factory farms to climate change and pandemic risk.[72] Their initiatives include exploring pioneering and novel technologies, such as cellular agriculture, which is meat produced without the exploitation of animals. In the meantime, while we sit and wait for technology to catch up, by limiting our animal-derived calories, by sticking to buying our meat and eggs from more traditional methods and by shunning the produce of factory farms we can help to move things in the right direction.

Contemplative Ethical Eating Key Points

- Where applicable, favour products certified with ethical credentials.
- Limit animal-derived calories to 10 per cent.
- Avoid meat and eggs from factory farms, favouring produce from game or animals reared more humanely.
- Seafood may be included, favouring molluscs and sustainably sourced fish.
- Limit dairy and stick to organic milk and butter from grass-fed cows.
- Better still, opt for plant-based milks.

Chapter 8

Contemplating Food and Togetherness

Growing up, my parents would make sure that most of our family meals were enjoyed together at the kitchen table. We'd enjoy what, usually, was prepared by my mother and we'd chat about the day's events. Sometimes the radio would be on and some of the topics would stimulate further conversation. Often on Sundays, we'd migrate to the dining room and feast on a traditional British Sunday lunch. On occasions, we'd enjoy the company of guests, or we'd visit friends' homes for a meal. Food would also be the centrepiece for meals with others for birthdays and celebrations. There was a lot of eating with the company of others.

Sadly, as I entered adulthood and attempted to navigate surviving an ever-increasing busy 21st century, meals that were consumed while sitting at a table with family and friends have become less frequent occasions. Worse: many an evening meal is devoured while sitting in front of the TV. Other than in the company of guests, sadly, there are now too few occasions when my wife and I dine at a table. I fear that my own behaviour is illustrative of that of many households in modern times. Indeed, according to recent research, almost a third of British adults eat alone most or all of the time.*[1] Fortunately, reflecting on my misguided behaviour has nudged me to correct my

* Based on a survey of 8,000 people for Sainsbury's by Oxford Economics and the National Centre for Social Research.

wayward conduct and urge my wife and I to sit together at a table occasionally.

Eating as a community has been enjoyed by most human societies for hundreds of millennia. Consequently, anthropologists feel it's probably an evolved strategy that reinforces social togetherness with positive effects on our wellbeing via our underlying biology. Socialising with food at the centre has been integral to the eating practices of all cultures with many sociologists and biologists arguing that eating in the company of others has important health benefits comparable to the nutritional value of the foods that we eat. Being part of a community has been shown to be integral to a happier, more active and more content population.[2] In this chapter, we'll explore some of the influences of our modern lifestyle on the way we eat and how togetherness influences our health, longevity and mental wellbeing and the underlying biology.

A Time to Eat

The phrase 'three square meals a day' originated from 18th-century naval voyages. Meals were served on a square wooden tray in order to avoid it sliding around on a rocking ship.[3] The time of day meals have been consumed and the names given to feeding times have changed over the course of history, although it's likely that the concept of three meals a day started with early agrarian societies. Meals have, however, been a traditional practice across multiple cultures for millennia. Some repasts have remained fashionable, such as the lengthy Sunday lunch or a celebratory dinner out at a restaurant. On the flipside, the way some meals are eaten has changed drastically, such as the way many of us wolf down our breakfasts as we're running out the door. The vernacular continues to change, too, depending on where you live and your social background, often causing confusion, such as the time I arranged to meet a friend for 'dinner'; he expected me at around noon, whereas I was set for an evening event.

'Breakfast', the first meal of the day and a term coined in the Middle Ages,[4] makes sense for hard-working farmers. Leaving home early with a full belly helped those having to endure a long and hard day's labour, especially when the next meal may not have been until late in the afternoon. Breakfast in most regions was – and still is in many rural areas in developing countries – based on starches like rice, corn or oats, foods now known to be energy-dense slow-release carbohydrates: ideal for arduous labour. Lunch, however, has not always been the essential meal that many people today believe it to be. Prior to the 18th century, dinner was consumed mid-to-late afternoon during hours where there was still daylight. As time went on, the increasing demands on the workforce meant lengthier working days, and breakfast was consumed earlier and dinner pushed further into the evening. This mandated a new meal to fill the gap.[5] Although often a lighter meal in Western cultures, in other regions lunch can be the main meal of the day.

The meal that's been subject to the most diurnal alteration is dinner. The time that we have consumed the meal referred to as 'dinner' has bounced around from midday until late in the evening throughout history, especially in relation to working times. In Victorian England, for example, the working-class main meal was served around midday due to the demands of domestic servants preventing them from dining at night.[6] This explains why some people in the UK still refer to 'dinner breaks' at work. For the affluent, dinner was taken in the early evening. The term 'supper' refers to the meal that's typically the last of the day. Supper may, indeed, be synonymous with dinner, or can be a light pre-bed snack.

The terminology, types of foods consumed and number of meals has shifted throughout history and continues to vary widely across the modern world. But the one thing they all have in common is that they are traditionally enjoyed in the company of others.

Community's Overwhelming Grip

The claim that our pre-agrarian ancestors didn't have three meals a day is evidenced by observing the eating behaviour of modern hunter-gatherers, of which few remain. Such societies are present in different regions of the world, such as the Amazon, Indonesia, the Congo Jungle and the East African Rift Valley. The Hadza tribe are one such society that resides on the Tanzanian savannah, situated in the Rift Valley, the very area where the origins of the Homo lineage can be traced to, around 2.5 to 3 million years ago.[7] The Hadza tribe are hunters and foragers and, although their culture will have altered over time, they provide a useful representation of the eating habits of our ancestors prior to 12 millennia ago. Stephan Guyenet, in *The Hungry Brain*, provides a detailed description of how members of the Hadza hunt, forage and share their food.* He describes a tribesman who follows the tracks of a kudu and relates his activities as he encounters potential food sources along the way. He also outlines the excursions of a Hadza woman as she forages for fruit and tubers. As the hunter comes across food during his trek, he'll gorge on as much as he can comfortably consume along the way – a not insignificant amount. He would then bring as much as he's able to carry back to his fellow tribe folk. The behaviour of the foraging women would be similar, devouring considerable quantities of fruits before carrying the reminder home. Guyenet describes their gorging behaviour as

---

* Guyenet provides a detailed narrative of typical hunting and foraging expeditions of Hadza tribe members which is well worth a read. Here, I've merely provided a very brief summary from his account.

'gluttonous'. Although they consume vast amounts in one go, they do so in the knowledge that their next feed may be some time away.[8] Upon the return of the hunters and foragers, other tribe folk prepare and cook the acquisitions. Later in the evening, all members of the tribe enjoy a large meal together. These highly social occasions involve all the tribe being involved in the meal preparation in some way, providing a sense of fulfilment for all comrades. Andrew Jenkinson, in *Why We Eat (Too Much)*, provides similar observations in relation to the Hadza after spending time with several families. He notes that 'Learning from the lives of hunter-gatherer populations is essential in order to understand who we are now, and how we reacted to our changed environment'.[9]

The fact that all cultures, including hunter-gatherers, enjoy social eating implies that the consumption of food in the company of others is an evolved strategy that's provided survival advantages both to individuals and groups. Functioning as a group promotes the release of feel-good chemicals like serotonin, oxytocin and endorphin and a reduction of stress-associated cortisol, providing neuroscientific and endocrinological rationale to further validate the claim. Similar behaviours have been observed in other primates who display social acts that benefit their mutual wellbeing, such as grooming each other and playing together. Humans are social mammals for whom functioning as a community and eating together has numerous benefits.

Family Mealtimes

The COVID-19 lockdowns forced many people to experience a greater degree of social isolation, which, of course, meant not being able to eat meals in the direct company of friends. For some, working from home meant time was less of a precious resource, giving more opportunity to enjoy meals with their family. However, despite only sparse data being available, one US study revealed some disappointing – but not wholly surprising – news. Researchers interviewed parents between December 2020 and February 2021, and the results suggested that their children's handheld electronic device usage had increased during mealtimes, and that this was associated with increases in body weight.[10] Sadly, it seems, kids are craving distractions which take them away from family conversation.

Dining in the company of others brings value for everyone around the table. Not only do we make more favourable food choices, research has highlighted numerous physical, psychological and social benefits. Children and adolescents who participate in regular family meals tend to consume more fruits and veg, are less likely to consume unhealthy foods and sugary drinks and have lower rates of obesity.[11] This could be because mealtimes help to establish healthy eating patterns and portion control. Children and teenagers who dine regularly with their families have lower rates of depression and anxiety and exhibit higher self-esteem with a stronger sense of identity and belonging.[12] Such occasions seem to nurture a sense of attachment and support contributing to improved emotional wellbeing. Regular family meals can contribute to stronger family bonds, providing a setting for family members to connect, share their experiences and offer support to one another. Sitting at the table and eating as a family can lead to healthier, happier and more connected families.

The Roseto Effect

Single parents, those who work awkward hours and families who for whatever reason have limited opportunities to eat in a traditional family unit needn't despair. Any occasion that involves sharing food will have merit even if it's not with blood relatives. Breaking bread with friends boosts the physical and mental wellbeing for all involved. Moreover, these benefits aren't limited to mealtimes, they extend to any social activity, and are particularly evident in those with a community-based lifestyle. The physiological and psychological benefits from this are worth discussing and are illustrated by a particularly interesting phenomenon.

In her book *Growing Young: How Friendship, Optimism and Kindness Can Help You Live to 100*, Marta Zaraska argues that the benefits of living a community-focused life and socialising have profound effects on health and longevity. She makes reference to 'The Roseto Effect'. In 1960, the local physician observed that Roseto, a small town in Pennsylvania, USA, had barely any cases of heart disease in locals under 65. This was highly unusual when the USA was experiencing rapidly growing rates of cardiovascular disease, and Americans in the nearby communities were suffering heart attacks at far higher levels. Indeed, Roseto enjoyed mortality rates 30 to 35 per cent lower than surrounding areas. When scientists looked into the physician's observation, they concluded that the effect wasn't due to the Roseto gene pool, and nor was there a link to their food habits: 'The Rosetans loved sugary treats, cooked with lard, and enjoyed sausages – 41 per cent of their calories came from fat', Zaraska explains. They drank wine and hard liquor, and they 'smoked and worked gruelling hours at a quarry or at a local factory'. Moreover, many of the Roseto locals were obese.

Years later, research revealed a surprising conclusion. Zaraska continues, 'Roseto's unusual healthiness was due to outstanding sociality, which has roots in the town's history.' Immigrants from the Italian town of Roseto Valfortore founded the town in the late 19th century. When they settled in Roseto, they continued their traditional jovial behaviours following Italian traditions, caring for friends and family members and lived in multigenerational homes. Their lifestyle was very community-focused, and included much celebration with frequent gatherings in their gardens with lots of food and wine. The town had a number of clubs for various sports and pastimes such as fishing, hunting and a library. The locals were very neighbourly and looked after the upkeep of the town, keeping it clean and tidy and planting flowers.

It wasn't long later, however, before the younger adults started to swap their community-focused way of life for the 'American dream', chasing hedonic, luxurious lifestyles and trying to get ahead of the Joneses. In 1971, Zaraska lamented, the town reported its first heart attack in a person under 45 and, by the late 1970s, Roseto's health outcomes resembled that of any other US town.[13] The Roseto effect is a useful illustration of how a community-based lifestyle can have a positive effect on our health. Enjoying good relationships with others and adopting a social mindset might well positively impact on our physiologies to a level comparable to consuming a good diet. How is it, then, that being in the company of others influences our health? Fortunately, there's considerable interest in this area, including research in relation to eating with others.

The Calming of Others

A powerful reminder of the benefits of conviviality on our health, remnants of the Roseto's community spirit can still be observed in Italy today. Social eating remains an unvarying characteristic of the people living on the Italian peninsula where the love for communal social action pervades every aspect of the society. Indeed, tangible evidence of *il piacere di stare insieme*, which means 'joy of being together', presents itself all too frequently. To this day, pictures of large Italian families gathering around a dining table abound in Italian TV commercials, luring viewers to buy whatever product is being advertised to experience the same sense of delight and togetherness. But that captivating, heart-warming image has to be squared with the growing malaise of modern workers. In most Western countries, including Italy, the introduction of a lifestyle with long working hours and increased social isolation has led to soaring levels of stress.

Of course, there are evolutionary reasons why we need stress. Some degree of stress is key to survival. But too much can be harmful and has been linked to cardiovascular disease, some cancers, depression, chronic pain, gut issues and obesity.[14]

From a scientific perspective, stress prompts our body to release cortisol. Produced by the adrenal gland, this hormone plays a key role in an automatic response that prepares the body for times when we have to accelerate our metabolic activity in response to external conditions, like when we encounter a threat or flee from it. Once the alarm to release cortisol has sounded, your body becomes mobilised and ready for action. In evolutionary terms, this mechanism allowed our hunter-gatherer ancestors to escape imminent and potentially life-threatening dangers. Today, this system works more like the prolonged, high-pitched noise of a police siren wailing off down the street: the noise is there but with time, we learn to ignore it. Or at least until the continual low-level disquiet starts to cause physical wear and tear. According to the Health and Safety Executive (HSE) of the United Kingdom, between 2022 and 2023 stress, depression or anxiety accounted for 54 per cent of all work-related ill health cases.[15] This is because constant exposure of our tissues to cortisol ravages our immune system, elevates blood pressure and hijacks glucose metabolism, leaving you more susceptible to a range of cardiovascular maladies.[16] But that's not all: cortisol can also have particularly dangerous effects on your waistline by interfering with your hypothalamus to increase the palatability of highly caloric food, a phenomenon we call 'stress eating'.[17]

There are a number of ways that we can reduce the amount of cortisol we release and one area has particular relevance. Over the past decade or so, considerable evidence has emerged demonstrating that the number and quality of close friendships we have has a significant impact on our cortisol levels. Studies on primates have shown that social connectivity dampens cortisol levels both during highly stressful situations and everyday activities.[18] In one such study, the scientists concluded that 'regular and repeated, everyday affiliations' have the potential to normalise our stress response, boosting our physical and mental health.[19] It goes without saying that the regularity of sharing a meal with someone certainly falls into this category and, as we shall see below, evidence is mounting of its extraordinary benefits.

The Comfort of Others

Our motivation to eat does not only stem from visceral, round-the-clock satiety signals: emotional and psychological processes are also important factors. These processes might also explain why we like eating with others. Food is often prepared to show affection to loved ones, develop durable bonds with friends or show hospitality to strangers. Consequently, food doesn't merely represent a means to satiety, it's also an instrument to build trust among people. In some situations, it may even function as a way to express our emotions – think about when you received a cake from your co-worker after you helped her to meet a gut-wrenching deadline. In fact, the amount of food you eat might even vary depending on who you are with,[20] showing that eating behaviour and interpersonal processes are intrinsically intertwined. This is being increasingly recognised by scientists all over the world as their attention shifts from the intrapersonal effects of food to the interpersonal processes. Essentially, how food connects and impacts on yourself and others rather than the impact it has on you.

There are several reasons for the growing appetite for this field. To name one, social eating is a fundamental form of cooperation that has characterised our species for thousands of years. When you think about it, it is hard to imagine how any progress in our history, societal organisation and cognitive development could have happened in the absence of food sharing. Romans would spend up to eight hours dining with friends during their *convivia* (Latin for 'living together') banquets, where the host would strive to impress his guests with elaborate dishes and diverse forms of entertainment.[21] One obvious benefit of these convivia is that they contributed to build and reinforce mutual alliances among people, increasing the size and quality of one's social network. This, in turn, can have profound effects on someone's health and wellbeing, not least because a wider social network means more emotional and social support. Studies have shown that friendship and social support can greatly decrease your risk of illness later in life and even death.[22] We also know that enjoying pastimes with others increases our sense of bonding towards those with whom we engage in those activities. The exact mechanisms behind this remain murky but a lot of this may have something to do with what scientists call 'behavioural synchrony', a concept that can be applied to any activity where people engage in synchronised, cooperative behaviour like dancing, singing or playing team sports. Piles of studies show that synchronised activities have a formidable soothing effect on people by turning down their pain and kindling feelings of closeness.[23] Coordinating behaviours with a complete stranger has been shown to increase empathic feelings towards that person, possibly as much as with a close friend.[24] Of course, strictly speaking, social eating can't be considered a synchronised behaviour – no one in their right mind would sync their chewing to their dinner-mate's – but it has, however, been suggested that eating with others might help us live longer and happier lives while feeling increased engagement with our communities.

The Garden of Eden

Spearheading this line of research is Professor Robin Dunbar from the University of Oxford, who, in 2010, recruited over 15 athletes from the Oxford University Boat Club to dissect the health benefits of synchronised physical activity.[25] Rowing is a great example of cooperative behaviour as athletes are continuously required to work synchronously to maximise their efforts. Dunbar's findings revealed that team players can tolerate twice as much pain as those who train alone, effectively increasing their 'pain threshold'. Rowers are required to work in synchrony, which in turn dampens their feelings of pain by ramping up the levels of a group of chemicals in their bodies known as endorphins. Endorphins are released by our brain in response to physical pain, and they are also linked with pleasurable sensations like laughing or exercising. What is of particular interest here is that research has shown that endorphins are also involved in the regulation of eating behaviours. Because eating promotes both social bonding and feelings of pleasure, the discovery that eating and bonding behaviours share one common cogwheel in the brain has prompted scientists to explore whether this group of chemicals could be behind the positive effect of social eating on health and wellbeing.

In search of answers, Dunbar's team ploughed through large amounts of data on social eating generated by one of the UK's largest educational charities. The Eden Project, situated in Cornwall, England, has sprawling biomes giving rise to the largest indoor rainforest in the world. In an attempt to encourage people to gather and connect to their communities, this charity launched 'The Big Lunch' in 2009 – where swarms of food lovers were invited to take to the streets to share a meal with their neighbours. Sniffing the opportunity to sample a large number of bread-breaking individuals, Dunbar embarked on a ground-breaking study to find out how neighbourly nattering and food sharing improves someone's wellbeing and social network. His work, summarised in 2017, concluded that eating with someone not only generates more bonded relationships, but also enhances one's sense of contentedness and wellbeing.[26] Where do endorphins come in? Although this study did not look at levels of endorphins, researchers think these chemicals are behind the uplifting effects of dining with others. The research on this issue is in its infancy, but the prevailing theory is that the frequent release of these natural painkillers helps to dampen feelings of stress or anxiety, ultimately strengthening our immune systems and firing up our natural defences to illness.

A Problem Shared

Another aspect of social eating has long captured the fascination of brain nerds all over the world. As the sharing of food resources can increase feelings of closeness to others, some scientists believe this process is key to maintaining an empathic connection towards others. Empathy is increasingly being recognised as a key component of social cognition and cooperative behaviour, as it allows us to make sense of and respond appropriately to others' emotions. Although some mystery remains around the connection between food sharing and empathy in humans, two lines of evidence seem to point in this direction. First, for most people, witnessing a friend in discomfort over a break-up or some misfortune at work is likely to cause empathic concern and personal distress. Under these circumstances, showing supportive behaviour towards your friend – for instance, by offering food – has the overall effect of both decreasing stress for the recipient and lessening empathic distress for yourself. However, there are a few cases where this type of behaviour goes awry. Differences in empathic responses have been observed in a number of psychiatric disorders including schizophrenia, autism spectrum disorder, borderline personality disorder and depression, and, more recently, evidence has been provided for individuals with eating disorders. Recent studies have suggested that social and emotional difficulties may be a key factor in developing these conditions. In other words, people who fail to understand and identify the emotional states of others may be at greater risk of developing some sort of eating disorder.[27] Interpersonal difficulties often play an important role not only in causing the illness but also in maintaining it, meaning that individuals with eating disorders may be less likely to interact with others or to fully share their feelings.[28] Evidence seems to suggest that feeling more connected to others may reduce the risk of psychiatric difficulties, in turn protecting someone from developing eating disorders later in life. This hypothesis is supported by extensive evidence in the UK, indicating that the risk of depression in vulnerable individuals is reduced by almost a quarter when people join a social group such as a sports club or hobby group.[29] And this is where eating with others comes in: surveys on visits to pubs or social evening dinners seem to converge on the idea that social eating, like other joint activities, has evolved as a mechanism to facilitate social bonding, and food offering functions as a powerful empathic response to another person's emotional state.[30]

The second line of evidence comes from past research on the negative effects of social stress on feelings of empathy. When two individuals are unknown to one another, their ability to feel empathy is somewhat inhibited by negative emotions like anxiety or fear. Bumping into a stranger may be perceived as a threat, particularly in socially anxious individuals who often prefer to avoid social interactions to avoid stressful stimuli. For these individuals, walking into a crowded space or joining a party requires enormous courage and causes their bodies to release large amounts of hormones that make them focus on the perceived 'threats'. This behaviour impairs their ability to read other's feelings, creating a barrier to empathy in the presence of strangers. How do these individuals overcome this barrier? It turns out that sharing an experience as simple as playing a video game with a stranger can reduce social stress and move us closer to that person – in essence, moving from the 'stranger zone' to the 'friend zone'.[31] This line of research shows that basic strategies to reduce social stress can break down communication silos and pave the way to empathic behaviours. Among those strategies, social eating functions as a powerful enhancer of empathic behaviour by fostering stronger relationships and trust among people. As it turns out, offering a stranger a biscuit at a business function may be a lot more than a mere act of generosity.

Lessons from Anorexics

Research points to another hormone: oxytocin, a chemical messenger synthesised in the hypothalamus also known for its effects on social bonding, lactation and maternal behaviour.[32] More recently, much attention has been drawn to the fact that this 'love' chemical also plays a central role in the neural circuits controlling appetite, body weight and food intake.[33]

Oxytocin is a hormone that helps entwine appetite with social bonding. Experiments in rodents have even demonstrated that pumping oxytocin directly into the brain can override those food reward mechanisms with the overall effect of silencing your urge to eat certain foods.[34] What happens when the levels of this hormone reach unusually low levels? This is where a large chunk of research on oxytocin and eating behaviour currently sits. Given its role in sparking interpersonal interactions, scientists have begun to ponder whether reduced oxytocin may explain bonding difficulties in patients with eating disorders. A key feature of patients with eating disorders is their reluctance to engage in social behaviours, and although this trait is well documented in the scientific literature, a flurry of potential mechanisms have been suggested to explain it. Among those, the idea that low levels of oxytocin may predispose individuals to emotional difficulties is currently riding high. The evidence starts with neonates, where bonding activity takes place around breastfeeding. Breastfeeding is one of the several activities that ensures emotional attachment in human infants towards their mothers and could be viewed as a distant forebear of social eating in adult life. The act of eating together is a ritual that brings benefits beyond the immediate satisfaction of basic human needs – i.e. satiating your hunger comes with the added bonus of forming connections with those around you. From a scientific perspective, the manifestation of these repeated 'attachment' behaviours are crucial in childhood as they gradually become engrained in the memory system, allowing the child to develop internal working models of how to interact with the rest of the world. This is why some scientists now believe that when this attachment behaviour is repressed or disrupted in early life, people may go on to develop eating disorders in adulthood. Oxytocin released during breastfeeding is thought to prop up early mother–child interactions by fostering sensitivity and synchrony. Although evidence in this area is still meagre, some scientists contend that adverse attachment events in early life may lead to dwindling oxytocin levels and increased vulnerability to eating disorders. Why? As far as the theory goes, adverse attachment experiences – such as a failure to establish an early bond with your mother – in early life may trigger an increased stress response, paving the way to long-term emotional instability and eating difficulties later in life.[35]

Although the literature on the link between oxytocin, early maternal behaviour and eating disorders is sparse, studies on individuals with anorexia nervosa (AN) are starting to paint a compelling picture. AN is a condition characterised by self-starvation and distorted body image and is concerningly common among adolescent females as well as a growing number of males. Individuals with anorexia constantly worry about how much body fat they have, despite being worryingly thin, and obsess about how much food they consume. People with AN often experience a profound sense of alienation from their community which leads them to develop anti-social behaviours. Some of these difficulties arise from problems in their attentional processing and manifest in excessive attention to negative body shape stimuli – not surprisingly, teasing, body shaming and bullying are suggested to trigger the onset of this illness.[36] Because of the devastating effects of this condition, scientists have long sought to understand the mechanisms underlying decreased motivation to eat in these individuals. Not only do people with AN disengage with food cues, but they perceive them as threat-related stimuli. Unlike healthy individuals, anorexics display increased activation of a region of the brain called the anterior cingulate when looking at images of food. This region is implicated in evaluating the emotional valence of objects in the environment as well as attentional control. The increased activation of the anterior cingulate in anorexia suggests that looking at food cues triggers anxiety in these individuals as opposed to positive emotions.[37] One study also found that, while consuming a meal releases copious amounts of dopamine in the brain of healthy individuals, increased dopamine release might be a source of anxiety for anorexics, rather than a source of pleasure.[38] Because humans are evolutionarily wired to avoid potentially threatening stimuli, anorexic patients seek to minimise exposure to these stimuli by directing their attention away from them. It might be for this reason that people with anorexia alienate themselves from food and the social setting in which it's consumed.

As it turns out, people with AN have abnormally low levels of oxytocin,[39] to the extent that the lower the levels of oxytocin in some patients, the more severe their eating disorder.[40] However, among its other functions, oxytocin is known to decrease cortisol levels, which, in anorexics, tend to be particularly elevated.[41] Cortisol is produced in response to stress, and lower levels of oxytocin in AN are thought to facilitate cortisol spikes in anorexics. As a result, an increasing amount of attention is being directed to the cortisol-dampening action of oxytocin in the hope that it may one day lead to new strategies to reduce anxiety and stress in individuals with AN, and subsequently alleviating the effects of their eating disorder.[42]

How does the diminished levels of oxytocin in anorexics translate to contemplative nutrition? The implication of the function of oxytocin in social cognition and eating behaviour provides scientific validation of the influence of social bonding and eating in the company of others on our wellbeing. It also serves to emphasise the importance of family meals on children's development. Studies in this area are ongoing and greater scrutiny on its mode of action, as well as the impact of intraindividual, interindividual and contextual factors, will be required in order to create a more robust picture. Nevertheless, these findings help to illustrate the benefits of interacting with others around food on our wellbeing.

Contemplating Eating with Others

My original précis for the contemplative nutrition regimen focused on optimal nutrition for physical and mental health, sustainable nutrition and ethical eating. Although I always appreciated the importance of social eating, I hadn't fully grasped the extent to which the benefits of eating with others extend beyond the obvious social merits. Sharing meals with friends and family is a key part of a healthy dietary strategy benefiting both our own wellbeing and that of others. Disappointingly, in the West, fewer and fewer of us are regularly enjoying our meals with others and this is having adverse health implications. In January 2021, the UK supermarket chain Sainsbury's polled 2,000 households and disclosed that only 28 per cent of UK families share the same meal each evening with one in five households preparing a totally different meal for different family members.[43] Consequently, emphasising the importance of social eating is crucial, and this is why it has its own contemplative nutrition pillar.

Another long-term influence of eating meals as a family is that the benefits transfer across generations. Parents who report eating a greater number of family meals per week while they were growing up also report that they now eat meals with their current family more frequently, as well as having a more regular meal routine pattern.[44] Indeed, in a culture of fast foods, it's not just the eating together that's lacking, food preparation is too. Researchers have observed that poor ability to cook and plan meals, as well as lack of time, were among the reasons that fewer parents spend time preparing food and opt for pre-packaged meals.[45] However, if you seldom have the opportunity to dine with others, enjoying other pastimes – playing sports, going to the gym, dancing, coffee mornings, playing cards at a bridge club or even going to the pub – will help your physical and emotional wellbeing, and might even prolong your life. It's the being among friends and family that counts the most.

Contemplative Social Nutrition Key Points

- If you can, sit at a table and enjoy a meal with family or friends three or four times each week.
- Try to have regular mealtimes.
- Whenever possible, share the process of meal preparation with others and vary your split of tasks: it helps teach new skills.
- If a friend or colleague is having a hard time, offering them food or making them a drink can make you both feel good.
- If you live alone and seldom have the opportunity to eat with others, prioritise socialising around other activities and find pastimes you enjoy.

Part III

Rethinking Eating

Chapter 9

Contemplating Contemplative Nutrition

We all want to enjoy what we eat. The pleasure we derive from the flavour, texture and subsequent postprandial satisfaction of what we put in our mouths is an emotion that we intrinsically pursue. Nowadays most of us live in societies where an abundance of different foods are readily available. For anyone living in such a society, the acquisition of a nutritious and varied diet is easy even if you have self-imposed dietary restrictions. Few vegans, for instance, report lack of culinary enjoyment. With this in mind, when choosing what to eat, we can factor in other considerations – like sustainability and ethics – into our decision-making. But any diet must, first and foremost, be one that optimally supports your own health. Your wellbeing has to be the number one priority and a diet must, at least, support good health. But, let's face it, in affluent societies – even for those on lower incomes – minimal effort is required to prepare meals that support a nutritionally rich diet; all that's needed is planning. Consequently, when it comes to making food choices, taking into account considerations such as animal welfare, land use and food-derived greenhouse gas emissions needn't be hard for most of us once we have access to all the information.

Contemplative Nutrition

Ideally, we should all incorporate a degree of rational decision-making when choosing what we eat rather than blindly adhering to food cues. In order to make sound judgements, having some understanding of the food-related issues outlined in this book will be useful. But the level of knowledge required to make better choices need only be a basic awareness of the key health, sustainability and ethical issues, and the tips provided in the last chapter will provide a summary of the main considerations, simplifying contemplative nutrition.

Just as our ancestors made practical considerations to acquire sustenance, we too can adopt a mindset that contemplates our options. Granted, the way our ancestors considered their nutrition was very different to what I'm suggesting. Even up until recent decades, the acquisition of sufficient nourishment was the primary consideration. These days, when presented with an abundance of choice, it's relatively easy for us to be mindful of other factors. Therefore, a sensible eating strategy should seek to address and provide solutions to the problems of the modern food system. This is contemplative nutrition. It is defined as follows:

Contemplative nutrition is a reasoned consideration invoking multiple perspectives of what you should include and what you should try to limit in your diet, prioritising important science- and moral-based issues, most notably those relating to nutrition and health, wellbeing, the environment and ethics.

Where possible, a contemplative diet prioritises considerations grounded in science and ethics over those based on emotions, feelings and beliefs. However, organoleptic (your senses' preferences), cultural, religious, social (what those around you eat) and habitual (what you're used to having) influences will still play a key part in your food choices, especially as these factors will impact what you deem to be ethical.

Mindfulness Eating

Just as psychologists are encouraging us to be more self-aware, contemplative nutrition involves us being more mindful about the foods we eat. A contemplative nutrition philosophy provides a desperately needed solution that will address the impact of our food system on issues relating to the physical and mental health of individuals, the climate emergency and the questionable farming systems that contribute to the unnecessary suffering to animals, while allowing sufficient production of affordable, convenient and enjoyable food that will satisfy human desires. To simultaneously acknowledge crucial modern health, ecological and ethical concerns individuals should try to make considered and reason-based decisions. If the way we view food doesn't change, we're likely to witness worsening levels of metabolic disease, mental wellbeing and ecological disasters that will ultimately affect all of humankind, animals and the environment.

Fortunately, a suitable dietary strategy can be realistic, sensible, straightforward and enjoyable. Importantly, contemplative nutrition need not involve prescriptive dietary regimens: it merely encourages reflection when selecting what to eat. Moreover, a contemplative diet for one individual might be quite different to that of someone else. The common feature is making food choices that are advantageous to the individual's physical and mental wellbeing while causing minimal unnecessary suffering to others – both humans and nonhumans – and adverse environmental impact.

Mindfulness eating shouldn't be confused with what is often referred to as 'intuitive eating'. Also known as 'mindful eating', intuitive eating is a simple idea that involves rejecting the dieting mentality. By recognising feelings of hunger and satiety, the strategy suggests you should allow yourself to have any food you want: no food is off limits. Indeed, some people find it a useful method to control what, when and how much they eat. However, as we've seen, appetite, hunger, cravings and satiety involve complex mechanisms with biological, psychological and social influences, and, consequently, food cues aren't always the most reliable guide. While it can be useful to recognise how you're feeling in response to food, that's only part of what I'm proposing. Mindfulness eating in the context of contemplative nutrition is actively considering the multiple perspectives – health, wellbeing, the environment, ethics and others – when it comes to choosing what to eat. Intuitive eating can play a role in mindfulness eating.

As humans we're equipped with the necessary cognitive and physiological flexibility to change our diets for the better. Hundreds of thousands of years of evolution have taught us how to successfully fine-tune our nutritional intake by leveraging the seemingly infinite amount of food sources that have accompanied our existence. This makes us adaptivores.

Adaptivores

Humans are complex biological systems and the realisation that we are adaptivores helps us embrace this often underappreciated fact. There is no one diet that nature has prescribed for us. What our ancestors ate varied dramatically over time and space, driven by what was available in their ecosystems at the time. Forged by uncertainty, humans have evolved the ability to survive and thrive on an impressive diversity of foods and with different proportions of plants and animals.

The term 'adaptivore' needs to be distinguished from 'omnivore'. An omnivore is a creature that possesses the ability to process nutrition from both plant and animal matter. However, an omnivore might only acquire the sustenance it requires from a fairly narrow range of foods. An adaptivore has the ability to glean nutrition from a huge range of foods and is able to adapt this ability under different conditions and circumstances. 'Adaptivore', of course, needs proper definition. First, however, we should look at what makes human food habits different from other animals. How is it that we can consume so many other organisms to successfully sustain ourselves? What's involved in being an adaptivore?

Optimal Foraging

This ability to acquire nutrition from a wide range of foods is a survival advantage that many other species lack. Cows, for instance, can only munch on grass, foliage, grains and some fruits,* and lions rely on their ability to catch prey for sustenance. Why then, if their physiologies limit their ability to process nutrition, are so many animals picky eaters? If you lived in a harsh environment with a scarce supply of food, why would you want to make things even harder for yourself by being finicky?

As Marvin Harris pointed out in *The Sacred Cow and the Abominable Pig: Riddles of Food and Culture*, very few species consume all potentially accessible food sources, despite them possessing alimentary tracts and metabolisms able to support the digestion and absorption of the valuable nutrients and calories available from these shunned foods.[1] Wolves, for example, seek out

---

* Modern industrial farming practices feed cattle with other forms of nourishment, such as oils and fish remains. But these are not foods they've evolved to eat.

and hunt only particular species, passing up on opportunities to consume many edible creatures present in their local habitat. Likewise, humans eschew potential food sources despite our ability to process nutrients from almost all the animals and a large amount of plants available to us. How many of you eat grasshoppers, caterpillars or other creepy-crawlies when you're feeling a little peckish and the bugs are at arm's reach? And I'm betting you've a strong aversion to eating your pet dog, despite his carcass being a rich source of protein, fat and vitamins! Of course, in these times of plenty, we can freely pick and choose what we want to avoid because alternative foods are just a short saunter to the fridge door. But only recently has there been an abundance of food. The struggle our ancestors had to endure for their next meal might seem incomprehensible today, despite many modern populations having to endure similar arduous ordeals. Why, then, are convenient sources of protein and calories avoided in preference of potential acquisitions that involve considerably greater levels of work hunting and foraging?

The reason why might be explained by a concept known as 'optimal foraging theory', which describes how creatures adapt to the potential food sources that are available.[2] The theory states that evolution favours the selection of behavioural strategies that maximise an animal's net energy intake when compared to time spent foraging, hunting, handling and preparing food. For example, the potentially huge calorie and protein gain from hunting a big animal is likely to be worthwhile for a wolf pack, even if the energy cost, time spent and risk are high. Although in many respects it might be easier to hunt smaller animals, such as rabbits, there's still some cost for a meagre net calorie benefit per hour of work. For similar reasons, certain human tribes typically shun many potential food sources, such as insects and snakes. However, if larger and more fruitful food sources become really scarce, then the smaller, lower net nutrition foods are more likely to be sought out.*

This helps to explain why, until recently, humans didn't eat green leafy vegetables in any great quantities. It just wasn't worth the hassle. Leafy greens provide very little energy and protein despite containing vitamins, minerals and fibre. Today's intensive agricultural methods and diminished soil health have led to reduced levels of micronutrients meaning it's more beneficial for us to spend time gobbling down salads, especially as foraging for veggies means little more effort than throwing a few additional items in your supermarket trolley. Previously, when we were getting sufficient roughage and key micronutrients elsewhere, there was less need to bother with greens.

Opportunistic Omnivores

Few animal species share the fruits (or meats) of their forages, at least in any appreciable amounts. Certainly, humans are unique in that we're the only primates who share their acquisitions with others to any notable degree and doing so has provided significant evolutionary advantages. While some members of our social group were out hunting for dinner others were gathering berries and tubers. This division of labour helped to ensure a stable, dependable source of food for everyone in the tribe when the hunters had an unsuccessful day. Next time your partner goes shopping for ingredients to cook a meal for friends or when you pull your smartphone from your pocket to order a pizza, consider how this incredibly flexible and adaptable strategy continues to be highly advantageous for our species. Our

---

* Optimal foraging theory involves calculations to evaluate the costs and benefits of hunting and foraging. Although the theory has been challenged,[3] the concept helps to explain the exclusion of readily available foods in animals' diets.

predisposition to sharing our food has been a major component to the development of human altruistic behaviour. Sharing is caring.

Our willingness to share our catchings and our flexibility to select or shun potential sources of sustenance helps to demonstrate that humans are adaptivores. Adaptations have allowed our ancestors to flourish amid highly challenging environments through evolving the ability to acquire nutrition from a vast array of different foods. Evolutionary adaptations have furnished us with the ability to process proteins, fats, carbs, fibre, vitamins and minerals from numerous animal and plant food sources, permitting us to readily adjust the foods that make up our diets when mandated by external pressures.

These adaptations began to really take shape around seven million years ago when our hominin ancestors split from our chimpanzee and bonobo cousins. From early members of our genus, around 2.3 million years ago, through our ancestors acquiring the ability to manipulate fire and use tools, right up until recent millennia, multiple adaptations in human populations have enabled the evolution of physiological traits best suited to the varied environments that we've frequented. As observed by the anthropologist Herman Pontzer in his book *Burn: The Misunderstood Science of Metabolism*, 'diet and metabolism are such strong evolutionary drivers that we can adapt to almost anything we have to eat'.[4] There are numerous examples from archaeological and genetic records, such as the ability for adults to digest the milk sugar lactose: a mutation that happened in two separate populations independently.[5] Another is how genes that relate to high-fat diets are more common in populations whose diets are seafood-rich, such as in Inuit populations.[6] Or indigenous tribes living in Chile's Atacama Desert who have adapted to the naturally high levels of arsenic in their groundwater, with natural selection favouring gene variants that help speed up the clearance of arsenic from the body.[7] Pontzer also notes the importance of studying the data acquired over the past several decades from multiple hunter-gatherer, pastoralist and traditional farming cultures. All populations consume a mixture of plants, fish and animal foods, though at considerably varying amounts; their ancestors were forced to adapt to what was available.[8] Of course, a large number of the plants present in our environment aren't suitable for human consumption. Despite us being surrounded by grass, leaves and foliage – sources of nutrition enjoyed by many other species – their metabolic cost of digestion is so high that it's not worth it. Adaptation is bi-directional: just as we're able to acquire nutrition from a vast range of foods, we've adapted to *not* being able to process nutrients from costly flora. It benefited our ancestors to favour more nutrient-dense fruits, tubers and honey.

Humans are an adaptable and flexible species with the ability to eat a lot of what's potentially available to us. In the words of Herman Pontzer, 'we are opportunistic omnivores': there is no singular, natural human diet.[9] Our ability to garner sustenance from food, however, isn't just about physically evolved mechanisms, it equally involves psychological, emotional, social and cultural influences.

Cultural Impulses

As well as selection processes that have benefited our lineage occurring at the level of the gene, cultural memes have also been responsible for adaptations. Memes are cultural ideas, traditions and behaviours that evolve in a similar fashion to genes. The memes that are advantageous for populations are selected for and subsequently spread across populations and societies. They include dietary preferences that have been built into cultural food laws and customs, and often these practices have involved religious influence. These socially propelled mechanisms have been key in helping humans decide what to eat.

To understand how human culture influences what we eat, first let's consider what we mean by 'culture'. Etymologically, the word is directly linked to the acquisition of food. Its early origins include 'the tilling of land; act of preparing the earth for crops'; 'a cultivating, agriculture'; 'to tend, guard; to till, cultivate'; and 'the cultivation or rearing of a crop, act of promoting growth in plants'.[10] Despite many people thinking of culture as the opposite of 'biology' – as in 'nature vs nurture' – human culture lies at the core of human experience; the two are inextricably linked.[11]

Massimo Montanari, in *Food Is Culture*, describes culture in relation to what we eat: 'What we call culture takes its place where tradition and innovation intersect'.[12] Tradition is itself an innovation: arguably the most successful one in history. Culture, through slow incremental processes linked to evolution, has helped humans reconcile our needs with what nature has had to offer during times when food was harder to acquire. From this, food culture has evolved into so much more.

Montanari describes the early dietetic relationship with culinary preparation in the medieval era, and how the health qualities of food – or, at least, what they were perceived to be – were linked to the pleasure derived from food. In Europe, much of these ideas stemmed from the writings of the Roman physician Galenus who, in the second century, expressed the pairing of four elements: 'hot' and 'cold' and 'dry' and 'moist', linking the balance between them to the optimal health of an individual. Galenus' influence on culinary choice continued through the premodern era and provides a rationale behind some obscure food combinations we eat today. Providing a 'balancing' act of cuisine served to make 'hot' foods 'colder', 'dry' foods more 'moist' and vice versa. The foods, of course, are not actually hot, cold, dry or moist, but their classification according to Galenic tradition serves to pair foods with the result of broadening the variety of foods eaten,[13] helping to ensure that all nutritional needs were met during times when food choice was limited. This has led to the pairing of many less obvious food choices that are eaten today, such as ham and melon, roast pork and apple sauce, and Christmas turkey with cranberry sauce. This tradition wasn't only associated with European culture: examples include sweet and sour in Chinese cooking, sweetened korma Indian dishes and honey-crusted pigeon in Moroccan cuisine. It's even possible that this medieval tradition provides justification for the highly controversial ham and pineapple pizza topping.

Many historical cultural food-processing methods are still practised today. In rural South America, for example, some communities still experience pellagra, a condition associated with a deficiency of niacin (vitamin B3). Symptoms of pellagra include inflamed skin, mouth sores, diarrhoea and dementia. In these regions, maize is a staple food for many. Some grains, such as maize, need to be processed through nixtamalisation, which involves the grain being soaked and cooked in limewater* and then hulled. The tough husks are broken down by the alkaline solution and this makes the niacin bioavailable for absorption in the intestines. Unfortunately, in some remote South American cultures where nixtamalisation isn't practised, cases of pellagra can still be found. Nixtamalisation is a crucial process in the production of tortillas, hominy and other maize-based breads, foods traditional to many South American cultures, and where these are regular staples, pellagra has been prevented.[14] In other regions of the world, white flour is fortified with niacin, which, along with access to a more varied range of foods, helps to prevent pellagra.

The impact of food culture is also greater the younger we are: we've become conditioned to favour foods based on our childhood preferences. This is illustrated when a child in a certain culture acquires particular culinary tastes and desires and declines foods prepared in

---

* That's the mineral lime, not the fruit.

the style of another culture. For example, children from tropical areas who've been raised on certain spicy foods embedded in their local traditions adapt to tolerate these strongly seasoned foods. The consumption of spicy foods is associated with cultures rooted in hotter regions and adults raised in more temperate zones, unaccustomed to foods prepared with such pungent spices, may find them unpleasant or even intolerable. The addition of some spices to foods has an antimicrobial effect helping to prevent food spoilage and to provide some protection to the human digestive system as a way of possibly mitigating the risk of bacterial or parasitic illness. This adaptation has been vital for those who reside in regions with higher incidence of pathogenic diseases. These stomach-led culinary customs may have helped to ward off food poisoning.[15]

These sorts of evolutionary adaptations are an example of something known as 'Darwinian gastronomy', culinary behaviours that have provided a reproductive advantage.[16] The historical importance of the antimicrobial effects of spices in hotter climates is part of a wider cultural influence. For example, spicy foods can trigger a temporary increase in metabolic rate giving a small rise in body temperature which, in turn, stimulates sweating, helping us to cool off in hotter temperatures.[17] Thus, spicy foods have two seemingly unrelated functions for those living in hot climates: not only do they stave off bacterial infections, they also help to keep people cool. Moreover, hot weather suppresses our appetite and adding spices to food stimulates our biological responses to food cues by making meals more flavoursome and interesting, increasing our appetite for them. The effects of cultural tradition validate the choice of ingredients and the characteristic flavours associated with each culinary cuisine has helped to prevent food from becoming boring.

In more temperate climates, the use of salt has emerged as a dominant preservation method as most microbes can't survive salty environments. In colder regions, food preservation is easier through natural refrigeration from the cold weather, and in hotter climates, traditionally meat was smoked and then dried in the heat to help preserve it. The salting of food has been practised for millennia, a discovery mainly attributed to the ancient Egyptians, though residents of the Shanxi province in China were fighting wars over salt reserves from the salt lake Yuncheng as early as 6000 BCE.[18] One of the reasons why the salting of food is still widespread today is that many human populations have evolved to favour salt flavours. This provides another example of Darwinian gastronomy. We taste salt and our brains are informed that the food is safe for us to eat.

These adaptive mechanisms that involve our physiologies, habits, cultural beliefs and emotional attachments to food help to explain, in part, why we have flavour preferences for certain foods and culinary customs. Reminded by our biologies, we remember which foods we ate were safe, nutritious and, above all, pleasurable.

Adaptivore Advantages

An adaptivore is an opportunistic omnivore that can readily adjust its diet to glean nutrition from a huge range of plants, animals, fungi and microorganisms, and this ability to adapt its eating habits occurs at the levels of the gene, the individual, its social group and its species. The definition of an adaptivore is as follows:

An adaptivore is an omnivorous animal who, through adaptations that occur – or have occurred – at the level of the individual organism, its social group and the species, possesses the ability to readily adapt to be able to acquire suitable nutrition from a wide range of plant, animal, fungi and microbial matter.

As adaptivores, we can acquire nutrition from a wide selection of foods available in multiple different environments. This ability is due to the countless combinations of physical and behavioural adaptations that have taken place by natural selection, social influences, cultural traditions and how we were raised. Moreover, what we eat also affects the microorganisms that reside symbiotically in our digestive systems. Not only do these microbes influence how we process food, but they can affect our emotions and behaviour, too. Consequently, adaptations can also occur at the level of the microbiome: as the bugs adapt, so do we.

Could it be the case that as long as what we consume is within certain sensible boundaries – i.e. we have a reasonably varied and balanced diet that supplies sufficient amounts of all essential nutrients and fibre and is not excessive in calories – no single dietary strategy is better or worse than another?* Although it's debatable, data suggests that the effects of the majority of the various diets – such as low-carb, paleo, vegan and low-fat – on metabolism are not as dissimilar as advocates of each would have us believe.[19] Does this mean that, as adaptivores, we can get away with eating nothing but hamburgers, cookies and chocolate? Unfortunately, the answer to this is, of course, highly unlikely. Constituents of food (such as sugar, salt and, maybe, saturated fat) as well as, possibly, the way some foods are put together mean that if we consume too much of these foods too regularly, we put ourselves at risk of chronic disease. And there's a limit to how many calories we can consume without suffering adverse health consequences. Plus, as we've seen, a good intake of fibre is important for optimal long-term health. For our ancestors, being an adaptivore ensured they survived long enough to pass on

---

* At least from the perspective of nutritional and energy value. Behaviourally, certain diets may work better for some people than others, due to reasons such as adherence.

their genes to their progeny, to raise their progeny and to provide support for their progeny's progeny. Although it might not provide the same merits living in the West of Plenty as it did in the past (or, indeed, for those in less affluent regions today), being an adaptivore was how our ancestors – right up until as recently as a few decades ago – flourished. Even though we can consume all those delicious high-fat, high-sugar, low-fibre, calorie-laden foods in the quantities many of us do, there is, sadly, no adaptive reason why their regular consumption would support us living healthily to a ripe old age. Simply put, from a metabolic perspective, we have a reduced need for our adaptivore advantages than our ancestors did. Therefore, in a modern food environment, being an adaptivore might be working against us. Or is it?

Being an adaptivore means that we have an advantage that's fundamental to contemplative nutrition. By utilising many of the new food technologies, we can adapt. We have the ability to continue to optimally and adequately nourish ourselves while simultaneously addressing the impact of our food choices on environmental destruction, the ethical concerns of the modern industrial food system and soaring rates of NCD. Our adaptivore advantage will be crucial for us as we deal with these existential challenges.

Non-Contemplation

There's something about the human psyche that means that, before anything else, we principally care about three things: what food tastes like, how much it costs and how convenient it is. Sure, some people alter their diets because they care about the environment and animal welfare, or they make food choices based on the perceived nutritional attributes of products. But for most people, their motivation stops at anything beyond pleasure, price and ease. Just look at the prevalence of junk-food-rich diets despite the fact that most people intuitively know such choices aren't going to be ideal for their long-term health. Fundamentally, most of us *don't* contemplate what we eat. In order to tackle environmental issues and the metabolic disease crisis we have to produce enjoyable and convenient alternatives that are available at affordable prices. What we choose has to be both beneficial for our health and have minimal environmental impact.

In *Meathooked*, Marta Zaraska states that we should learn from our ancestors and, 'instead of looking for a perfect and "natural" diet from the past, start looking for one that would be best for right here and right now'.[20] Although when it comes to what we should be eating there might be some disagreement on what 'best' might be, with an ever-growing population and worsening rates of chronic disease it seems clear that making more healthy and sustainable food choices necessitates utilising the learnings of nutrition science. This is where the advances endowed by food technology will help us tackle the problems of the food system. This is where food processing comes in.

Chapter 10

Contemplating Nutrition Science

Nutrition science, although a relatively new discipline, has origins that can be traced back to ancient Greece where the writings of Homer, Plato and Hippocrates acknowledged the role food plays in our health.[1] Yet, it wasn't until the late 18th century, following key advances in chemistry, that real progress occurred. One early discovery was in 1789 when French chemist Antoine Fourcroy recognised a distinct class of biological molecules, including albumin, gelatin and gluten[2] – but it wasn't until nearly 50 years later that a Swedish chemist, Jöns Jacob Berzelius first described these chemicals as proteins.[3] Similarly, the term 'vitamin' was coined in 1912,[4] but the first organic micronutrient was isolated in 1924.[5] Since then, nutrition science has exploded. Through extensive nutrition research, our knowledge of how food and nutrients are linked to deficiencies, disease risk, performance and cognition has expanded exponentially. However, despite this, we continue to be showered with a mishmash of cherry-picked and dietologically led inconsistencies, even when seemingly credible peer-reviewed research is cited. The resulting bombardment of opinions leaves the public confused, mistrustful of food policy, not knowing which way to turn and with little idea as to what sound dietetic advice looks like. This pessimistic realisation might leave us asking if it's even worth bothering with nutrition research.

Prior to nutrition science, humans didn't have the luxury of experts advising what's safe and good for our health. Our forebears relied on cultural practices; techniques passed down through generations afforded them the required skills to navigate what was safe and beneficial to consume. As adaptivores, our reliance on our sense of taste has been crucial for our evolution. Systems intrinsic to the many food cultures of the world have benefited us through our innate neophilia – the pleasure we derive from novelty – and neophobia – our comfort with the familiar. These features have allowed us to be generalists deriving satisfaction from both strategies. Although cultural traditions continue to influence our food choices, in the modern Western world our need to be reliant on them is considerably diminished. Consequently, these generalist attributes are being exploited by the myriad confusing food cues hurled at our biologies. In *The Omnivore's Dilemma*, the food journalist Michael Pollan notes that humans are 'anxious omnivores' and that we're struggling 'to find out what is wise to eat'. He points out that 'instead of relying on the accustomed wisdom of a cuisine, or even the wisdom of our senses, we rely on expert opinion, advertising, government food pyramid, and diet books, and we place our faith in science to sort out for us what culture once did with rather more success'.[6]

If deceiving our innate responses is a crime, then food science is guilty as charged. Should we, therefore, ignore nutritionists? Should dietitians be sacked? Should we be rounding up all the food scientists and shipping them off to concentration camps in remote regions of Siberia? The frustration that we feel from the conflicting misinformation might lead us to ignore the phenomenal benefits granted by nutrition science. The countless advantages afforded to us by nutrition research far outweigh the numerous controversies. Food science has enabled crucial discoveries that have improved the health of billions of people. Take Norman Borlaug, for example, who provided the assurance of sustenance to the most extreme and poverty-stricken regions of the world. The provision of vitamin supplements in developing nations has been invaluable, as illustrated by the fact that a daily vitamin A capsule has saved millions of child lives.[7] From shedding the pounds to packing on muscle, nutrition research informs us the strategies that are effective, helping to prevent us from getting lost in the persuasive dietology. And let's not forget that many of us are able to choose from a huge variety of foods every day, made possible through technical methods of food preservation and distribution. More crucially still, nutrition science will support us into the future: the continued learnings provided by ongoing research will be fundamental in helping us tackle the multiple health, ecological and ethical issues presented in previous chapters. Despite the numerous advantages provided by nutrition science and food technology, however, there remains much puristic criticism, especially when it comes to food processing.

Nourishing Innovations

While there are valid environmental, ethical and health concerns relating to our food system, and although we must acknowledge that food processing has undoubtedly contributed to many of these issues, the fact that nutrition science has helped us to mitigate a multitude of food-related issues is grounds for continued optimism. Food science and dietetics present us with opportunities that can assist us in finding solutions to the multiple crises we face. From food security to climate change and dwindling biodiversity, with the tools of nutrition science at our fingertips we have the potential to solve these problems. Although there are many issues that might be causing us increasing levels of concern, there are several exciting innovations that can help support the long-term physical and cognitive health of an ever-growing population. Let's look at some of these innovations.

Good Moral Objections?

The familiar term 'GMO-free' is used by marketers on food labels to promote products to those who wish to avoid genetically modified ingredients in their food. The number of people who object to the use of GMOs is surprisingly large: a recent report by the Pew Research Center found that half the US adult population is wary of them.[8] It's not uncommon to come across someone who's 'driven to red-faced anger at the idea that people are "messing with their food"', to quote the philosopher of social science Lee McIntyre.[9] In his 2021 book *How to Talk to a Science Denier*, McIntyre notes that, despite the huge public outcry against their use, no credible studies have shown any risk in consuming GMOs. Concerningly, this anti-GMO activism has been effective in slowing down the progress of GMOs, and, he points out, in Europe there hasn't been a single approval of a genetically modified crop for domestic cultivation in 20 years. Even less ardent sceptics remain cautious, fearing that, as GMOs aren't 'natural', having them present in our food chain, they feel, may lead to health problems down the road.

What exactly are GMOs? Genetically modified organisms are crops that have been altered at the molecular level to improve attributes such as the nutritional value, hardiness and resistance to weather, pollution, disease or pests, spoilage, or ease of farming. People have been consuming them since the mid-1990s following the development of the Flavr Savr tomato, which was modified to prevent spoilage.[10] The benefits of GMOs go beyond allowing industrial farmers to bump up their yields: they've been instrumental in improving the health of many people, including children, in less affluent parts of the world. An example is 'golden rice'. Rice is one of the most commonly consumed foods, with more than 3.5 billion people depending on it for more than 20 per cent of their daily calories.[11] When scientists cross-bred rice with a gene found in the daffodil, they massively increased the beta-carotene content of the grain.[12] Beta-carotene is a red pigment and a precursor to active vitamin A, and it gives this rice its 'golden' colour. Making one of the world's most consumed staples rich in this sight-protecting vitamin has helped to reduce the number of vitamin-A-deficient children in the world, half a million of whom become blind every year.[13] Moreover, golden rice is more drought-resistant than other strains: a huge boon to sustainable agriculture, helping to mitigate the effects of climate change.

Why do so many people object to these 'Frankenfoods', as GMO foods are pejoratively called? Due to the absence of data providing cause for concern in respect of health, the most likely explanation for their opposition is that it's born out of emotion. For many, GMOs are seen as not 'natural' and this invokes feelings of fear. But what do people mean when they say 'natural' anyway? Although GMOs involve the manipulation of an organism's DNA, humans have been indirectly manipulating the genes of their crops and livestock for at least 10,000 years.[14] Early farmers, through selective breeding, picked out their best crops and replanted them in order to continually improve their yields. Those crops that were less fruitful were discarded and, along with them, their genes. This meant that within a single generation, the gene pool was altered. Similarly, animals with the best traits were favoured and this is why today we milk cows daily, have larger pork-providing pigs and have chickens that lay at least an egg every day. The principles of selective breeding are how Norman Borlaug produced his higher-yielding grains in the 1960s. You don't see scores of people objecting to these many millennia-old practices. The science of genetic modification has merely simplified and accelerated the selection process.

Paradoxically, the advantages that GMOs bring with them should be appealing to the most extreme purists. Where approved, GMOs have helped to mitigate many of the issues that would otherwise concern environmentalists. These include using fewer 'unnatural' pesticides, reducing greenhouse gas emissions by 26 million tonnes and requiring less land to be farmed.[15] Some people have expressed concerns about the effects of GMOs on soil health. And these concerns aren't completely without merit, but they really only apply to intensively grown GMO crops where there's been little regard for the land.

Labelling foods that obviously contain no GMOs as 'GMO-free' may sound ridiculous, but due to the extent of the opposition, I can see why companies do it. The scientific consensus from decades of credible research is that GMOs are safe.[16] GMO technologies are valuable tools that benefit the food system. Unfortunately, those most likely to suffer from the lack of progress in GMO science caused by the anti-GMO movement are those in less privileged positions than those who are objecting. The principal losers are those with limited access to sufficient weather-hardy, pest-resistant crops, and who would otherwise be nourished with valuable calories and protein. In the words of Michael Shermer, 'GMOs are scientifically sound, nutritionally valuable and morally noble in helping humanity during a period of rising population'.[17]

Meatless and Cellular Ag

Since the early 1980s, vegetarians have been able to enjoy veggie burgers. Early recipes typically consisted of a few different vegetables compressed with mashed potato into a patty; nothing like what we expect to be served when we order one of today's plant-based burgers. While not unpleasant in taste, beyond their circular appearance, the original veggie burgers bore little resemblance to the texture and flavour of a real burger* and were a nutritionally inferior substitute to protein- and iron-rich meat. Fast-forward to the 2020s, and the

---

* I speak from experience. As a fussy teenager, I ate them under duress when served by my mother.

plant-based 'meat' market has exponentially boomed into an array of meatless sausages, mince, pulled 'pork', 'chicken' breasts, steaks, canned 'tuna' and the like, and global sales of the category are predicted to exceed $74 billion by 2030.[18] Ingredients include vegetable proteins, like pea and soya protein isolate, rapeseed (canola), sunflower or coconut oil for fats, methylcellulose for fibre, various constituents that give the products their meat-like organoleptic qualities, and key micronutrients, particularly those that are frequently lacking in vegan diets. Some companies have perfected their meat mimicry so much that it can be hard to distinguish their meatless products from their flesh-derived counterparts. Meatless products have become invaluable accessories in helping to reduce our consumption of meat.

As well as vegan-approved meatless 'meats', soon you'll be able to buy vegan-friendly burgers made out of actual meat. The first lab-grown meat burgers created by Dutch researcher Mark Post and his team, cost around $330,000 each and took two years to produce. These were consumed at a London press conference in 2013.[19] These products show considerable promise in helping us solve our current sustainability and ethical challenges. Through a process known as precision fermentation, animal-derived stem cells are placed in stainless-steel cultivators where they're fed nutrients and growth factors. After a few weeks of dividing and growing, the resulting product can be shaped into burgers, sausages, nuggets or mince. According to Believer Meats, the Israeli company that opened the world's first lab-grown meat factory in June 2021,[20] clean meat will provide the nutrition of conventional meat without harming animals, and with 80 per cent less greenhouse gas emissions, 99 per cent less land and 96 per cent less freshwater.[21] The company produces chicken pieces, chicken breast, meat pies, kebab skewers and burgers. Each patty, they claim, will cost less than $10.[22] While still expensive, that's a considerable drop from $330k. And, as economies of scale permit, prices are likely to be comparable to conventional meat and plant-based burgers within the decade.[23]

Similar cellular agriculture – or 'cellular ag' – is available for other animal proteins like dairy, egg and gelatin. One company, Perfect Day, are already selling their animal-free ice cream in American stores. Lab-grown egg-white powder has been developed to be available as soon as commercial hurdles have been navigated. And synthetic gelatin is already available for confectionery products.

The potential for the versatility of cellular ag is vast. For example, modern diets are typically deficient in omega-3 fats, and products could be created to have a higher amount of these essential nutrients or with boosted levels of key micronutrients. As the meats are created in sterile environments, food safety risks will be hugely reduced. With no need to give antibiotics to livestock, this will help to combat the problem of antibiotic resistance allowing us to conserve the majority of antibiotics for human use (globally, around 66 per cent of antibiotics used are for animals).[24] Whether these foods are approved for kosher or halal diets is yet to be confirmed, and there's some controversy as to whether these products will be certified as 'vegan'. As there's no need for animals to be harmed, there is support from some vegan bodies.[25] And the implications for animal welfare are enormous: each year, billions of animals need not be born into lives of suffering. Paul Shapiro, in his 2018 book *Clean Meat*, calls cellular ag 'the second domestication'. He feels it 'could be the biggest upheaval in how we produce food since the agricultural revolution some ten thousand years ago', and that it 'could just be the answer to some of the most pressing problems humanity faces as we move deeper into the twenty-first century'.[26] The CEO of meat giant Tyson Foods, Tom Hayes, agrees. In 2018, in a Bloomberg article he said, 'If we can grow the meat without the animal, why wouldn't we?'[27] Winston Churchill may have been off by a few decades when, in his 1931 essay 'Fifty Years Hence', he wrote, 'We shall escape the absurdity of growing a whole chicken in order to eat the breast or wing, by growing these parts separately under a suitable medium',[28] but his predictions were aptly intuitive.

At the time of writing, three countries have approved the sale of lab-grown meat for human consumption – Israel, Singapore and the USA[29] – and, in July 2024, the UK gave the go-ahead for lab-grown pet food.[30] However, somewhat predictably, and much like with GMOs, there's been pushback. Italy has banned the sale of cellular ag meat, and in May 2024, along with 13 other states, Florida banned it in an attempt to 'save our steaks', under the guise of protecting their 'proud traditions' through conjuring up images of tough ranchers on the American frontier.[31] But the truth is, this nostalgic image has already vanished. Through subsidies provided to industrial farms, not only are small farms becoming increasingly rare, but most of the country's meat is pumped full of antibiotics. This realisation throws into question objections to lab-grown meat being less 'natural'. Meats produced through cellular ag could be a huge tool for tackling the climate crisis. Hopefully other countries will give cellular ag products a chance to thrive in the market rather than flippantly dismissing their immense value.

Food from Thin Air

And if these new technologies aren't exciting enough, how about consuming protein produced from thin air? Could we be getting more from our air than just oxygen? This type of cellular ag involves a specialised process and a specific strain of microbe that's been developed by genetic modification. These microorganisms are fed tiny bubbles of carbon dioxide and water extracted from air, along with nutrients like nitrogen, calcium, phosphorus and potassium – i.e. those normally absorbed by plants from soil. The resulting product is dried into a powder with a protein content of more than

70 per cent, and this can be incorporated into recipes. All that carbon that our industrialised lifestyles are throwing up into the atmosphere might just be of some use after all! Solar Foods in Finland claim that their bioprocess that extracts carbon and water from air is 20 times more efficient than photosynthesis and 200 times more so than meat.[32] To quote their website, 'Harvesting single-cell protein from nature for food was not possible for hunter-gatherers 100,000 years ago, or the modern human race up until today. It's possible for humankind to now enjoy a new kind of harvest.'

Solar Foods sent me samples of their product Solein, which I tried on its own, mixed in water and as part of recipes. The brown-coloured, fine powder is mostly tasteless except for a mild nutty tinge. It mixes easily and is very versatile. Although it's maybe not 'plant-based' in the strictest sense, it's certainly free from animal products. I can see it being an affordable and, hence, commonly used protein source in the production of commercial foods and maybe you'll even have some appear in your kitchen cupboard in the not-too-distant future.

Complete Foods

Another sustainable food category has been experiencing considerable growth: complete foods. These powders, drinks, bars, instant meals and other easy-to-prepare food formats are nutritionally complete: i.e. they contain adequate amounts of all essential nutrients (protein, essential fatty acids, vitamins, minerals and fibre), beneficial levels of other key nutrients (including slow-release carbs, phytonutrients and monounsaturated fats) and controlled levels of food constituents that have been negatively linked to health (such as sugar and salt), and are designed to support optimum health and performance.

Full disclosure: as co-founder of Huel Limited – currently the global leader in the complete food space – and the nutritionist responsible for all Huel food products to date, I naturally have a bias here. Nevertheless, the huge growth of the complete food* category is undeniable: in a little over nine years, Huel has sold more than 400 million meals into more than 80 countries.[33] And Huel is just one of several brands: the entire complete food market is forecasted to reach $6.3 billion by 2027.[34] Of course, the quality of products varies between brands and many avoid the use of animal-derived ingredients. Complete foods are convenient, affordable and allow the consumer to be reassured that their meal contains good nutrition. What they otherwise might have eaten may have been a less-nutritious alternative. Moreover, they are sustainable nutrition with a low impact on animals and the environment and produce minimal food waste. Although complete foods are unlikely to be eaten as a meal shared with family or friends, they can be a key asset to a contemplative nutrition strategy. They provide a nutritious, sustainable alternative to the junk, high-carbon-emitting foods that many of us grab when we're busy. Depending on the brand, they provide valuable fibre, omega-3s, vitamins, minerals and phytonutrients that many people are otherwise missing out on. Think of complete foods as 'Plan B': useful for those occasions when making your own nutritious meals is troublesome.

Unprocessing Food Processing

Is nutrition science doing an adequate job of translating the nutritional benefits of food into good health and nutrition policy? Some

---

* Products are sometimes referred to as 'meal replacements'. However, the term 'complete foods' covers a wider range of products. Complete foods aren't a 'replacement' for other meals, in the same way you wouldn't say soup and bread is a 'replacement' for pizza and chips; you either have one or the other for a meal.

people don't think so. They feel that food policy is both misguided and misguiding. And they're not entirely incorrect. Take, for example, the controversy surrounding saturated fat recommendations discussed earlier: the evidence that saturates are harmful is far from straightforward and the fact that there remains a distinct lack of conclusive evidence of their danger leaves us wondering why they're a key focus of nutrition policy. Much criticism relates to concerns that too much emphasis has been given to the nutritional content of food, rather than focusing on foods and dietary patterns. In his 2007 bestseller *In Defence of Food: The Myth of Nutrition and the Pleasures of Eating*, Michael Pollan introduced popular culture to the term 'food reductionism',[35] a concept that refers to the failure to consider food as food; rather, we reduce it to being considered in respect of its nutrient value. He claims that the value humans ascribe from food is greater than the sum of its constituent parts.

The person initially responsible for highlighting the issue of food reductionism was the Australian sociologist of food science and politics author, Gyorgy Scrinis, who attacks the concept of a nutrient-led view of food far more aggressively. Scrinis feels that describing what we eat in respect of its nutritional value is itself a significant health issue. After coining the term 'nutritionism' in a 2002 essay entitled 'Sorry Marg',*[36] he later, in 2013, devoted an entire book to the issue. In *Nutritionism: The Science and Politics of Dietary Advice*, he describes food reductionism as the 'dominant paradigm of nutrition science and dietary advice', and that it's primarily characterised by 'a nutritionally reductive approach to food'.[37] He states that nutritionism's characteristics include decontextualisation, simplification, fragmentation, exaggeration and determinism with respect to the role of nutrients. Scrinis's critique extends to both nutrition science

---

* Scrinis is referring to margarine, which he views as the ultimate nutritionism product.

and food policy. He's highly critical of official dietary guidelines that ascribe numerical values to recommended nutrient intakes rather than being based on food or dietary advice.

Both Pollan and Scrinis describe nutritionism as an ideology. In outlining the reductionist perspective, Pollan adds that 'the widely shared but unexamined assumption is that the key to understanding food is indeed the nutrient', and he sums up the mindset with: 'Foods are essentially the sum of their nutrient parts'.[38] Scrinis goes further and defines nutrition reductionism as referring to both the reductive focus 'on the nutrient level and the reductive interpretation of the role of nutrients in bodily health'.[39] In *In Defence of Food*, Pollan provides a broad perspective and makes some valid objections about a food reductionist mindset and the way we view food, encouraging people to readopt more traditional methods of preparing food. To some extent I agree with him: the value that food provides certainly does tell us that it is more than the sum of its nutrient parts. Moreover, the number of occasions when people take time to prepare their meals are concerningly few. However, that doesn't mean we shouldn't also be embracing the knowledge provided by nutrition science as understanding the food we eat is crucial for tackling the multiple problems of the modern food system.

Gyorgy Scrinis, in *Nutritionism*, articulates his concerns way more fastidiously. To illustrate, he refers to the change in the way trans fats were viewed – i.e. not only was their use previously permitted, but actively encouraged, yet, as we've seen, they're now known to be hazardous to health.[40] His worry is that there may be other ingredients in foods that, while currently permitted, may later be revealed to be similarly dangerous. Scrinis, although complimentary to Michael Pollan on acknowledging the term, is otherwise highly critical, accusing him of being reductive as he references individual nutrients in order to describe the healthfulness of particular foods.[41] In defence of Pollan he merely incorporates nutrient-focused explanations in order to help illustrate his point: after all, the strength of any argument is dependent on supportive evidence, which, in this case, is the connection between consumption of certain nutrients and disease biomarkers and outcomes. Moreover, Pollan's narrative presents a useful perspective of food: one based on tradition. Scrinis acknowledges that we should be utilising the skills of nutrition science but in a different way. He feels that a nutrition science study of food should look at dietary patterns and particular foods and compare them to biological markers and health outcomes in order to draw better public health and nutrition policy. As an alternative to the current reductive approach to food policy, Scrinis suggests one that informs consumers by describing the degree of processing: 'Defining and categorising food quality in terms of methods of primary production and processing, rather than in terms of nutrient composition, can be the basis of alternative dietary guidelines, nutrition education strategies, and food and nutrition policies and regulations'.[42] Much of Scrinis's argument targets processed food, and while he makes valid objections, his critique misses key issues fundamental to contemplative nutrition. Naturally, we all should be concerned about some of the ways food is processed, but I challenge some criticisms of food reductionism. Before we get to this, however, let's look at what people mean by 'processed food'.

A Processed Mindset

When people hear the term 'processed food', they might think of pre-made meals served in plastic trays ready to be nuked in the microwave for a few short minutes, junk-food snacks laden with nasty additives or modern variations of traditional foods with an ingredient list the length of a short novel. While some processed foods are certainly unfavourable, many people have an overly maligned viewpoint of them. This is often unwarranted.

Food processing is the transformation of agricultural products into food, of one form of food into other forms or 'the basic preparation of foods, the alteration of a food product into another form, and preservation and packaging techniques'.[43] Our food has been subjected to some level of processing for thousands of years. Members of our species were processing grains by grinding them between two stones – the process we call 'milling' – as far back as 30,000 years ago.[44] Indeed, as the application of heat to food is essentially a process, one could argue that we've been 'processing' our food for hundreds of thousands of years. Food processing, therefore, has been integral to human evolution. The first food processing in the modern sense appears to have been in 1809, when Nicolas Appert invented the hermetic bottling technique to preserve food for French troops. Appert's technologies contributed to preservation methods such as canning, invented by Peter Durand just a year later.[45]

Processing encompasses all aspects of food preparation, cooking and preservation. Most people, however, are aware that there are different degrees of processing. It's unlikely that they think of milk, cheese, hummus and packaged mixed salad when they refer to processed foods. More likely, they're thinking of foods and ingredients that have been processed to a far greater degree than more basic and traditional methods: things like cookies, fast-food burgers and confectionery. Nonetheless, the term 'processed food' does seem to have become excessively reviled and hearing it is often associated with negative perceptions. Cultural food processing has been crucial to human evolution and if our ancestors hadn't discovered ways of processing ingredients, not only would we not be enjoying many of the foods we today find gratifying, but we wouldn't be enjoying our lives in the way we do either. Having a plentiful food supply frees up our time so we don't have to hunt, gather or farm, time we can better spend doing other things whether for our own pleasure, the benefit of others or even working towards improving humanity's future.

These advantages have been afforded to us by both traditional and modern processing techniques.

To address the vagueness of what's meant by processed food, there have been numerous attempts to classify foods in respect of their degree of processing. In his 2000 book *The Unofficial Guide to Smart Nutrition*, Ross Hume Hall proposed a four-level ranking system for the types of processing foods have been subjected to. Foods were ranked based on the degree of processing, the amount of non-nutritional chemicals added to foods and the amount of fibre present in the end product.[46] Hall's classification has limited real-world validity. For example, Hall rated fresh fruit higher than frozen, which, in turn, he ranked higher than dried and canned, respectively. This fails to account for key factors like nutritional quality and the necessity of particular processes to distribute food. Often the levels of key nutrients in frozen produce may be greater since it's typically picked and immediately frozen preserving its nutritional value, rather than 'fresh' fruit and veg that may take several days before it lands in shoppers' baskets.

What about 'ultra-processed food'? This is a term that carries tones of dread in the voices of those who utter it and arouses fear when we see it written. Just a few years ago, the term was barely on anyone's radar, yet it has found its way into the common vernacular in much of the English-speaking world. Such is the extent of the angst it instils, you'd be forgiven for contemplating the fate of the human race if we continue ingesting these substances. Maybe, however, we should pause and consider if the anti-ultra-processed-food hype is truly as existential as some make it out to be. Are we right to fear all ultra-processed food? What's caused the explosion in debate?

The generally accepted definition of ultra-processed food (UPF) falls under a food categorisation system known as the NOVA classification.[47] The term appears to have first been used in 2009 by Carlos Monteiro,[48] a professor of nutrition and public health at the University of São Paulo, who went on to lead the team who compiled the NOVA system. It's rapidly grown in use since and has been popularised by high-profile nutrition academics on popular podcasts and social media and following the publication of the book *Ultra-Processed People: Why Do We All Eat Stuff That Isn't Food … and Why Can't We Stop?* in 2023 by Chris van Tulleken.[49] NOVA is a convenient classification system that categorises food according to the extent and purpose of processing, rather than in terms of the nutrients.[50] There are four groups, summarised as follows:

Group 1: Unprocessed or minimally processed foods

Edible parts of plants, animals, fungi and algae after separation from nature. Processes are limited to removal of inedible or unwanted parts, or methods like drying, grinding, roasting, boiling, fermentation, pasteurisation, refrigeration, chilling and freezing. These processes are designed to preserve natural foods, to make them suitable for storage, or to make them safe, edible or more pleasant to consume.

Group 2: Processed culinary ingredients

Ingredients such as oils, butter, sugar and salt are substances derived from nature by simple processes to make durable products suitable for use in recipes to prepare, season and cook Group 1 foods and to make them varied and enjoyable. They are not meant to be consumed by themselves, only in combination with Group 1 foods.

Group 3: Processed foods

Foods such as bottled vegetables, canned fish, fruits in syrup, cheeses and freshly made breads are made essentially by adding salt, oil, sugar or other substances from Group 2 to Group 1 foods. Additional processes include various preservation or cooking methods, and, in the case of breads and cheese, non-alcoholic fermentation. Most have two or three ingredients and are recognisable as modified versions of Group 1 foods, edible by themselves or, more usually, in combination with other foods. The purpose of processing is to increase the durability and shelf life of Group 1 foods or to modify or enhance their sensory qualities.

Group 4: Ultra-processed foods

UPF are formulations made mostly from substances derived from foods and additives, with little, if any, intact Group 1 food. They include soft drinks, packaged snacks, reconstituted meat products and pre-prepared frozen dishes. Ingredients may include sugars, oils, fats or salt, but also other foodstuffs that have been directly extracted from foods, such as casein, lactose, whey and gluten. Group 4 items are derived from further processing of food constituents, such as hydrogenated oils, hydrolysed proteins, maltodextrin and high-fructose corn syrup. Food may include additives like preservatives, antioxidants and stabilisers, or those used to imitate or enhance the sensory qualities of foods, such as colours, flavours, sweeteners and processing aids. A multitude of processes are used to combine the ingredients into the final product including those with no domestic equivalents, such as hydrogenation, extrusion and moulding. The overall purpose of ultra-processing is to create branded, convenient, attractive, hyper-palatable and highly profitable food products designed to displace other food groups.

NOVA set out to move the attention away from the nutricentric perspective and to instead consider food in respect of how it's produced. Rather than viewing food as its constituent nutrients, focusing on how it's prepared offers a new way of looking at what we eat. And this is not without merit. The value humans ascribe to the food we eat certainly is greater than the sum of its constituent parts, as demonstrated by the way food values have been integral to every human culture. Although, on the surface, the NOVA classification seems reasonable, does it really let consumers know if a food is healthy or, conversely, something with potentially damaging attributes?

Hall, Scrinis, Pollan, Monteiro and van Tulleken all share a desire to be less reliant on the nutricentric way we view food. I'm totally on board with this: the way food policy is devised and the way the majority of 21st-century Westerners consider food has indeed been overly reductive. My major gripe with NOVA, however, is that placing the food we eat into four simple groups based solely on the degree of processing is also reductive: it's just reductive in a different way. One could argue NOVA is more simplistic than the nutricentric view.

When it comes to whether UPF consumption increases the risk of NCD, the data is compelling: diets habitually high in foods that fall under NOVA Class 4 – i.e. ultra-processed food – are associated with greater risk of metabolic disease and obesity. High UPF consumption has been linked to cardiometabolic disease, stroke, type 2 diabetes, depression, irritable bowel syndrome, functional dyspepsia, cancer, frailty and all-cause mortality in adults, and associated with metabolic syndrome in adolescents and dyslipidemia in children.[51] The epidemiology can't be ignored. Van Tulleken covers this extensively in *Ultra-Processed People*.[52]

Regardless of the fact that the data linking high UPF consumption and poor health is compelling, there nevertheless remains a risk of bias, even when controlling for confounding factors. Those testing the link between high UPF consumption and metabolic disease will probably have their hypotheses validated because the majority of the UPF tested will be low in fibre, protein, essential fats and micronutrients, and high in sugar, salt and saturates. As well as this, it might be that those who eat copious amounts of UPF consume less nutritious fruit, veg, cereals, pulses and seafood. Moreover, as, on the whole, UPF is relatively cheap, it could be that people who consume it in large quantities are more likely to have a lower income, which itself is a risk factor for poor health. Could UPF consumption be a proxy for poverty? As unhealthy behaviours go hand in hand, UPF consumption could signify an unhealthy lifestyle: smoking, lower levels of physical activity and/or high alcohol consumption.[53] Despite attempting to adjust for these variables, there's no getting away from the fact that the UPF consumed by subjects in the trials will likely have a poorer nutritional profile irrespective of how the foods are processed. Of course, the hyperpalatability and energy density of high-UPF diets may mean more is consumed, but calorie for calorie this research bias is hard to fully mitigate.

Nevertheless, it is highly likely that people whose diets include a substantial proportion of mass-produced, hyperpalatable food in the form of UPF have a higher NCD risk even after accounting for these potential confounders. So, why, then, am I whinging about NOVA and the use of the term 'ultra-processed food'?

An Open Mindset

Science is the continual pursuit of knowledge in order to further our understanding, and dietetics and food technology are progressive disciplines that allow us to know more and to understand more. Without utilising the tools of nutrition science to explore what makes up our food and using our knowledge of its constituent parts, we wouldn't be able to reap the advantages of the multiple technologies that provide sustenance to those who otherwise have limited access to food. Moreover, food science allows those with higher incomes to enjoy a vast range of convenient foods. While we should accept many of the criticisms of the nutritionism paradigm – as many modern NCDs are indeed linked to overconsumption and Western dietary patterns – the puristic hardline anti-reductionist approach is too dismissive of nutrition science. Those of us who reside in modern

Western societies are extremely privileged. We have, presented before us, a huge range of foods, many of which we simply wouldn't have access to if it wasn't for processing technologies. More crucially, in less economically developed nations, methods of food processing have allowed exponentially growing populations to be adequately nourished.

Let's look at bread: for centuries, the only ingredients required for baking were flour, water, yeast and salt. However, a loaf would go stale within a day or two. Traditional households were large and a mother would bake a fresh loaf each day, confident that the whole thing would be devoured by her hungry family before bedtime. Now view the food label of a commercial granary loaf and you'll see it typically contains a long list of ingredients. For example, the list may read 'wheat flour, water, malted wheat flakes, malted barley flour, toasted wheat, toasted rye, wheat protein, yeast, salt, vinegar, soya flour, sugar, emulsifiers, barley fibre, antioxidants' and fortified with several vitamins. Many of the ingredients allow the product to have a longer shelf life and increase its durability for transport, making it more accessible to a wider number of people who don't have to possess bread-making skills. Of course, the ingredients also aid the taste, texture and versatility of the bread, making it more enjoyable. An unappetising loaf in a world of plenty is unlikely to be eaten, and bread is a crucial source of nutrition for many people. As we've seen, most of us aren't consuming enough fibre, so by ensuring bread doesn't go off quickly and by making it more appetising, people are more likely to consume fibre-rich varieties, and fortifying with key vitamins, such as thiamin, niacin and folic acid, and iron, helps to ensure people have a sufficient intake of these nutrients that are too often consumed in inadequate amounts. Moreover, as more of a loaf is likely to be eaten, less is thrown away, which helps to reduce food waste.

I agree that the majority of the foods that fall under NOVA's UPF classification should be consumed only in moderation. But the UPF definition is a sweeping generalisation, and this calls its effectiveness into question. Numerous nutritious foods are unduly vilified solely because they fall under this 'ultra-processed' banner, a definition which is, after all, just the opinion of a group of researchers and their disciples. An overly simplistic system of categorisation that demonises foods is profoundly unscientific – ideological, even. A 2022 report was similarly critical, accusing NOVA of bias, claiming that it 'classifies foods according to the assumed "purpose" for which they have been designed and produced. This approach introduces a subjective (perhaps ideological) bias in the food classification process that should be, on the contrary, as independently objective as possible.' The authors contended that the classification 'suffers from a lack of biological plausibility so the assertion that ultra-processed foods are intrinsically unhealthful is largely unproven.'[54]

Indeed, for many foods, the classification is unclear. For example, canned vegetables could be in either Group 1 or 3 depending on what they're canned in. Tomato purée – a versatile ingredient that packs an antioxidant punch – isn't either a clear Group 2 or 3, and coconut fat is listed in both 2 and 3. In *Ultra-Processed People*, Chris van Tulleken concedes that some foods fall at the margins of the UPF definition. He uses canned baked beans in tomato sauce – a much adored British staple – as a useful example, pointing out that they're a healthy, affordable food, and that some varieties may fall into NOVA class 3 and others into NOVA 4 due to the ingredients. He admits that, in this instance, 'we meet the limitations of NOVA, a system designed to look at dietary patterns rather than to evaluate individual foods', and that 'there is almost certainly a spectrum of UPF, yet exactly how or whether any one product will be harmful is impossible to tell because we don't just eat one food – we eat a range of foods.'[55]

Advocates of NOVA feel that along with the list of ingredients, other factors are integral to the UPF definition, including how a food is labelled and marketed, and its hyperpalatability. Van Tulleken asserts: 'if I'm struggling with whether to call a food UPF, then it probably is UPF'. He feels that if a product has been developed in a way that promotes its overconsumption, then it's a UPF.[56] On the surface, this makes sense. However, the fact that there is a debate about whether or not some foods fall under the UPF definition makes me question the concept of pigeonholing when it comes to branding a food as 'unhealthy'. What about foods that fall under the UPF banner whose flavour profile doesn't encourage overconsumption? What about products containing one or two so-called UPF ingredients that are high in fibre and protein, rich in key micronutrients, and low in sugar, salt and calories? What about foods that will be crucial to help us combat the increasing environmental pressures on the global food system?

The fact that the way in which a food is marketed is part of the UPF classification seems silly. While we should be questioning the rationale behind some nutrition and health claims, especially as they're used to mask unfavourable attributes of some products, why should the mere presence of a marketing claim be an indication that a food is a UPF and, consequently, unhealthy? Weirdly, there seems to be a blanket demonisation of breakfast cereals by NOVA proponents, even though many cereals have nutritional qualities. Take Shredded Wheat – a 100 per cent wholewheat product that's a nutritious inclusion to any breakfast, particularly because it's consumed with milk – which wouldn't be categorised as NOVA 4 by even the most ardent NOVA aficionado. Should Shredded Wheat not be marketed using a health claim? Surely, doing so helps to promote these nutritious gems over sugar-laden alternatives.

It's easy to forget that we're fortunate enough to be in the position to be able to criticise the use of UPF. The ready access to a vast choice of food that the West has been enjoying for decades is what many in lower-income nations are now justifiably desiring. A fast-paced world where people don't have the time – or the will – to prepare meals is one in which they rely on convenience. Even within higher-GDP countries, like the UK and US, higher UPF intake has been associated with those on a lower income.[57] The anti-UPF fearmongering serves to worsen how those who can afford little in the way of minimally processed produce feel about their disadvantaged situation.

Fundamentally, what concerns me is the flippant use of the term 'ultra-processed food', the negative perceptions and the emotion it carries with it, and with the anti-ultra-processing dogma becoming more widespread, my concerns have intensified. Humanity faces immense challenges: our future is dependent on the advances of nutrition science. We have more than eight billion mouths to feed, food-system-derived greenhouse gas emissions to reduce, food waste to mitigate, and inhumane and unsustainable agricultural practices to address, while simultaneously allowing our fast-moving world to continue at its ever-rapid pace. A dynamic society means we can enjoy many comforts. Access to convenient sustenance allows us to work and spend quality time with our friends and families, and allows society to develop innovations that benefit others and help to improve our future. We desire access to an array of foods grown far from where we live, provided safely, in good condition and with a long shelf life. A food system that embraces convenience has been fundamental to all the societal improvements that we continue to prosper from.

Granted, our food system faces challenges in ways far removed from those that humanity has previously experienced, including modern diseases related to overconsumption and poor dietary choice. And for this reason, now, more than ever, we must rely on food science to provide sustainable nutrition. In the same way as our early agrarian forebears granted the benefits of selectively breeding for sustenance, we look to key food technologies like cellular ag, plant-based meats and complete foods. This heightened pressure for a more sustainable food system tells us that processing technologies are integral to 21st-century eating. Innovative and cutting-edge practices enable us to produce delicious and nutritious food and drink at scale and at affordable prices. Processing techniques allow us to buy raw ingredients from farmers and make safe food via distribution channels to remote regions, and to reduce food waste. Processing helps to support specific dietary requirements, such as plant-based proteins for vegans, and provides safe products for those with food intolerances. Importantly, processing has a role in helping people make healthier choices, such as manufacturers tweaking recipes to add more vegetables to sauces, reducing the sugar content and fortifying foods with specific micronutrients or fibre. Our ability to process food is hugely valuable and should be embraced.

It's clear that food is more than simply the sum of its nutrient parts. As well as this, foods contain beneficial non-nutrient factors such as phytonutrients found in plant foods. These beneficial components, which include antioxidants, may help to reduce the risk of many modern maladies such as CVD and some cancers. As food technology continues to discover new phytonutrients, the importance of nutrition science in human health is further highlighted. If the knowledge gleaned from food science adds nutritional and culinary benefits and allows greater accessibility, then the advantages are vast. The increasing pressure for our food system to be more sustainable forces us to consider the contribution of what we eat to greenhouse gas emissions, the ever-growing problem of food waste, land and water use and how we treat animals. Moreover, if meat can be kept fresher for longer, less is wasted. Exploiting an animal only to throw away its nutrition-rich carcass is both practically and morally irresponsible.

In fact, some UPFs might be beneficial for our health and reduce NCD risk. The results of a prospective cohort study published in *The*

Lancet in November 2023 offer a different take on the if-it's-UPF-it-causes-disease mindset. The study looked at nearly 270,000 participants from seven European countries. Participants were free of cancer, CVD and type 2 diabetes at recruitment. Foods and drinks consumed over the previous 12 months were assessed and classified according to the NOVA system. Subjects were followed up after a median of 11.2 years. The results? Aligned with previous research, higher UPF consumption was indeed associated with an increased risk of multimorbidity of cancer and cardiometabolic diseases. No surprise there. Among UPF subgroups, associations were most notable for animal-based products and artificially and sugar-sweetened beverages. Again, no real surprise. However, the results revealed that some categories of UPF breads and cereals or plant-based alternatives were neutral or negatively associated with risk.[58] These findings present a hefty challenge to the popular belief that all UPF are linked to negative health outcomes.* Clearly things are way more complex. The popular assumption that *all* UPF foods are linked to adverse health events is probably wrong. It seems that we can't simply tar all processed foods with the same UPF brush: other factors are more important than simply how much a food is processed. While the study's findings need to be replicated, they are helpful in that they demonstrate that subgroup analyses of UPF are necessary when exploring the association between UPF and NCD.

As well as this paper, another group of researchers has demonstrated that a diet that limits UPF isn't healthy by default and that the types of foods people eat matter more than the level of processing.

---

* Like all research, this study has its limitations, such as the NOVA classification was implemented on dietary data captured more than 20 years previously; potential changes in modifiable behaviours during follow-up, especially after the diagnosis of NCD, were not possible to account for; unmeasured confounders, such as family history of NCD, could have affected the results.

The team showed how it's possible to build a nutritious and high-quality menu that follows the US dietary guidelines where the majority of the calories (more than 90 per cent) come from foods classified as UPF.[59]

Unprocessing the Terminology

It's fair to say that NOVA and the other attempts to rank food by way of the degree of processing are themselves too reductive, merely more simplistic alternatives to the nutricentric view. Humans like to put things neatly into boxes. Here we have a box labelled 'UPF' with 'bad' rubber-stamped on it. This is too simplistic. We live in a highly nuanced world where nothing is black or white, least of all our hugely complex food system.

To highlight three key points:

1. Classifications such as NOVA are too literal and should not be relied upon.
2. Such classifications wrongly imply that the more a food is processed, the 'worse' it is.
3. Viewing processed foods with a negative mindset is extremely limiting and will, likely, lead you to unnecessarily miss out on some very nutritious foods.

To be clear, concerns regarding a large number of commonly eaten UPF must not be ignored. Quite the contrary: many UPF should be consumed with caution. Nor am I wholly dismissive of NOVA and the term 'ultra-processed'. NOVA has some usefulness for exploring the epidemiology of dietary patterns and disease, in the same way that we look at, say, the Mediterranean eating pattern or meat-rich diets and how each relates to NCD risk. But what I object to is high-profile academics and nutrition communicators pushing the term into everyday use as a means to assess the healthfulness and quality of food, and criticising anything that falls under the UPF banner. The sweeping generalisations of an anti-ultra-processed food mindset is dietological. A term that unnecessarily stigmatises potentially beneficial foods based on a simple categorisation without considering a food's qualities fails in its objective. NOVA and the term 'ultra-processed food' should be confined to academia.

Furthermore, despite its ubiquitousness, most experts argue that a much-needed standardised, formally accepted definition of 'ultra-processed food' has yet to be agreed. In a December 2022 paper published in the *British Journal of Nutrition*, Michael Gibney of the Institute of Food and Health, University College Dublin, flags NOVA's subjectivity bias and cautions against the recommendation that all UPF foods be avoided. His key concern centres on NOVA and public health messaging: 'If the degree and nature of processing of foods are to be considered as an important driver of public health nutrition, then some level of objectivity in the definition of highly processed foods is needed.' He points out that some types of bread and breakfast cereals are classed as UPF, yet they positively impact nutrient intake, and that further gains can be achieved through reformulation. In opposing the 'reformulation of foods on the grounds that one cannot make an unhealthy food (subjectively defined) healthy', NOVA ignores efforts to encourage reformulation of foods to be lower in salt, added sugars and fat. He asserts that if 'the degree and nature of processing are to be considered within the strategies of public health nutrition policies, a robust, objective, evidence-based definition must be devised and the criteria for considering a food as highly processed must first take account of that food's impact on population nutrient intake'.[60]

Indeed, Gibney is not alone in academic circles in highlighting the absence of an objective UPF definition.[61] By acknowledging that a lack of a definition is a concern, many academics, dietitians, nutritionists and food-industry professionals are implying that they don't know what the term actually means. Despite this, many continue to flippantly use the term when describing those processed foods that we should only be consuming in moderation. Why are experts – who ought to know better – continuing to casually use a phrase that they agree lacks a suitable definition?

Society seems intent on categorising things, and this includes the food we eat. Clearly, both the nutricentric perspective and the process-led classification of viewing foods are imperfect. A superior system may involve encompassing multiple attributes of food, i.e. key nutrients – the sugar content, the amount of protein, the presence of essential fats, its fibre content, the quantity of salt, and the levels of vitamins, minerals and certain phytonutrients – along with how and to what degree it's been processed. A better system still would additionally account for cultural, social and, crucially, environmental considerations. Simultaneously accounting for these multiple factors probably means that a public-friendly methodology is not possible. The intricacies of the global food system involving the complex structure of food, how it's produced, how it's consumed and the value we ascribe to it might mean that any attempt to categorise food is an impossible task.

Instead, maybe, we should look to a widely used term that already exists. 'Junk food' has been around for several decades and most people understand it to mean 'a pre-prepared or packaged, hyperpalatable, convenience food that's high in fat, sugar and/or calories, and typically low in fibre and micronutrients'. Granted, 'junk food' has no universally accepted definition – although, as we've seen, neither does 'ultra-processed food' – but this lack of a rigid definition is the very reason why 'junk food' works. The term also accounts for those highly palatable foods that are high in fat, sugar and/or calories and low in fibre that don't fall under the UPF banner, such as cakes, certain varieties of crisps (potato chips), fast-food kebabs, and fish and chips. In casual conversation, most people intuitively understand that something described as 'junk' is, by and large, likely not to carry desirable healthful attributes and should only be consumed in moderation. Moreover, it excludes those nutritious, fibre-rich gems that NOVA brands negatively as UPF, such as granary bread, whole-wheat breakfast cereals and baked beans.

There is much confusion over UPF and how the categorisation relates to the healthfulness of food. 'Junk food' doesn't cause the confusion that the term 'ultra-processed food' does, and in a complex world it is the very imprecision of language that allows it to be understood. In the words of the 20th-century Austrian philosopher Ludwig Wittgenstein, 'The limits of my language means the limits of my world'.[62]

Nutrition Transition

If food policy in its current guise is flawed, maybe we should also be looking towards dietary patterns rather than solely relying on a nutri-centric perspective in devising guidelines. It's also fair to attribute some blame to certain food technologies for much of the 20th- and 21st-century health and environmental issues. However, utilising alternative food technologies might be our only hope of digging us out of the many holes into which we've inserted ourselves. Rather than a blanket slating of UPF, let's ensure that the way in which we process a food enhances its nutritional quality, makes it more accessible and sustainable or aids its convenience. Whole ingredients may undergo limited processing in order to prolong the shelf life, boost organoleptic qualities and aid healthfulness by the addition of certain nutrients. At the same time, we could apply caution to processing techniques that appeal to consumers' hedonistic urges and contribute to overconsumption. While we're dependent on food technology in order to address the multiple challenges, we must ensure that the right food processing methods are used in the right way for the right reasons.

So, where are we now and what's next? In *Meathooked*, Marta Zaraska refers to five recognised stages of nutrition transition.[63] A society begins by collecting food, i.e. hunting and gathering, then transits to stage 2, famine, which involves agriculture and unreliable yields. In stage 3, famine recedes, agriculture improves and severe hunger becomes a thing of the past, but food remains simple. Later, societies that experience an industrial revolution enter stage 4, degenerative disease. It's claimed that stage 4 is where the West currently sits, i.e. consuming too much overly processed, sugar- and fat-laden foods. Many African, Asian and South American countries are in the process of transitioning from stage 3 to 4 as their disposable income increases. Some nutritionists predict a transition to stage 5: behavioural change, where it's prophesied that changing our behaviour will move us back to eating foods similar to those consumed by stage 1 societies, i.e. less meat and more fruits, veg and wholegrains. There are, however, problems with this. It's uncertain that stage 1 societies actually consumed many wholegrains: more likely their consumption only became significant through agriculture as societies transitioned through stages 2 and 3. And crucially, how are we going to feed more than 10 billion people without looking elsewhere? We need agriculture and we need food technology. There's definitely a lot wrong with stage 4. Could stage 5 be something else? Maybe stage 5 could involve utilising the benefits of technological advances to provide adequate, sustainable and optimal nutrition to billions of people. These might include complete foods, cellular ag, GMOs and insect protein, utilised alongside other exciting technologies that allow more productive growing of fruits, veg and mushrooms; innovations like ultraurban farming where vertical units allow shoppers to grow fresh produce at home or buy it directly in local stores.[64]

As the environmental journalist Mark Lynas points out, 'if we resisted anything but natural farming technology (circa 1960), we would need a plot of land the size of two South Americas to feed the planet'.[65]

Chapter 11

Contemplating What to Eat

I recently heard a comment: 'You can tell when we don't know much about something by the amount of books written on the subject.' There are few subjects where this is more relevant than nutrition. The multitude of food-related literature illustrates the enormity of human ignorance when it comes to what we should eat. When bombarded by conflicting information, it's little wonder that the state of nutrition knowledge is poor and food-related anxieties are exacerbated.

Anxious Adaptivores

I'm fortunate to have the opportunity to dine out a few times a month with family, friends or business associates. With few exceptions, these occasions are enjoyable experiences made so by the environments in which I dine, the company I'm with and the food I'm served. However, during a recent restaurant excursion with friends, I noticed something about my own behaviour: despite staring at the menu for several minutes, I was no closer to making a selection. Of course, I'm a little handicapped in that my options are restricted since I heed my own advice. Nevertheless, I had a realisation: my brain was telling me that it was imperative for me to make the right meal choice out of the four dishes that would suit my requirements and likely please my palate. For some reason, I was apprehensive about making the wrong choice, despite knowing that I would likely take pleasure in wolfing down any of the four. Why, then, did it seem so important that I make the optimal selection? And how would I be able to assess my choice anyway? Even if I had a gadget that could tangibly compare the degree of pleasure derived from one dish to another, prior to receiving the meal I had no knowledge of the chef's flair for each dish, nor would I ever have known if I'd made the best decision, since I could only pick one. My hesitation made little sense. I'd succumbed to the paradox of choice. In reality, all had a high likelihood of gratification. Moreover, if the only pleasure I sought from the eating-out experience was what could be attained through the flavour and texture of the meal, then I may as well have opted for the less risky approach: stay at home and make myself my favourite meal. No, it seemed, being waited on in a pleasant restaurant, in the company of good friends with amusing conversation wasn't, to my hedonic-seeking brain, sufficient and I sought to boost the experience as much as possible. And even if the dish I ordered turned out to not be that great, I'd probably be enjoying eating out again in a few days' time anyway. This was a depressing realisation.

As humans, we're cued to seek to maximise our perceived pleasure of future food choices. This stems from us being anxious and insecure creatures. Human insecurities are both innate and instinctual. Our insecure instincts apply to all the things we crave – warmth, sex, friendship, money, excitement, value – and are why we seek out pleasure. Our brains have evolved to attribute pleasure to those behaviours that were useful for our survival as primitive beings, such as warmth, safety, procreation, companionship and motivation. Thus, hedonism provides an antidote to our insecurities, driving us to continually pursue what we enjoy. In no other behaviour is this more apparent than our drive for sustenance and the pleasure we get from devouring food. These cravings extend beyond food's taste and texture and include seeking out value from foods that have been associated with our upbringing. This reminds me of an apt quote made by the late British-born Zen Buddhist philosopher Alan Watts in 1951: 'The animal tends to eat with his stomach, and the man with his brain. When the animal's stomach is full, he stops eating, but the man is never sure when to stop. When he has eaten as much as his belly can take, he still feels empty, he still feels the urge for gratification.' This, Watts claimed, is largely down to our anxiety about the uncertainties of a constant supply of food, so we eat as much as we can while we can. But also that, in an insecure world, pleasure is uncertain so 'immediate pleasure of eating must be exploited to the full, even though it does violence to the digestion.'[1] More than 70 years later, as we suffer the misgivings of meal-selfie Instagram, Watts' intuitions are even truer. The human condition mandates that we think we need more sustenance than we actually do. In the modern world of plenty where few humans alive ever fear being unable to enjoy their next meal – let alone be denied access to food at all – how do we 'uncondition' ourselves from excess while continuing to be gratified through pleasure and value with the food we seek?

From one perspective, it seems rational that we should be insecure about our food supply. After all, our ancestors didn't have the luxuries of financial security and supermarkets. They relied on their cravings to motivate them to vigorously pursue their next meal. Yet our hunter-gatherer ancestors weren't as anxious about their food as you may think, and it's probably only been since the development of agriculture around 12,000 years ago that our food anxieties escalated, following unreliable harvests resulting from drought, flood, pestilence and theft. Research with the present-day hunter-gatherer San tribe of the Kalahari – the oldest surviving tribe in Southern Africa – has demonstrated that their members make rational decisions when seeking food. The San have survived unforgiving arid conditions for tens of millennia without exploiting and exterminating the animals and plants on which they depend. During a drought, they have the foresight to consider the consequences of what would happen if they killed the last plant or animal of its kind. They spare members of the threatened species, tailor their conservation plans to the varied vulnerabilities of edible fauna and flora, and enforce these efforts against the constant temptation of poaching, such as *if I don't take these last few springboks, someone else will*. They ignore their immediate anxieties and hedonistic urges, giving way to the reasonable promise of future security. In a society built around reciprocity and sharing it's unthinkable for a San hunter not to share his hunt with an empty-handed bandmate, aware that hard times befall all and he may one day require the extended hand of a neighbour.[2]

Another concept that could be contributing to the way we view food might also help to explain why food anxieties differ between cultures. The phenomenon relates to the way people see themselves in their society, including how they view food. Modern Western industrialised nations tend to be societies based on a social theory known as 'individualism'. Individualism is a moral perspective where the emphasis is on the individual. In such cultures, independence, self-reliance and goal-oriented thinking are the focus. In contrast, 'collectivism' is a social perspective where the emphasis stresses relationships in the social group and community, and many eastern and African states tend to lean more this way.[3] Individualist cultures tend to be more 'me'-focused, whereas collectivist cultures are more 'we'-focused. Someone with an individualistic mindset might feel that, because they work hard, they've earned the right to buy and eat whatever food they want. This perspective places a lower priority on the needs of others and the wider environment. This I-deserve-it mentality tends not to acknowledge that, when compared to most of history and, indeed, a large number of people alive today, the fact that you're able to sit down and enjoy a flavoursome, nutritious meal is actually a rare privilege.

I'm in no way suggesting that collectivism is a superior social outlook to individualism. Quite the contrary: an individualistic perspective can motivate people to work harder and it may well have been a key contributor to the higher standards of living many Western countries have enjoyed. Nor am I suggesting that you should alter your mindset, even if it's possible to. As many of you reading this will, like me, probably have been raised in more individualistic cultures, intuitively the way you view your own needs and those of your immediate family will be prioritised over those of others. This is not without merit, of course. As I said earlier, you can't effectively take care of others unless you're in good health yourself. Moreover, there can be disturbing consequences when people de-prioritise their own wellbeing. Thinking in a me-first way is probably so ingrained as a result of our upbringing that it's the norm. But acknowledging that there is a different perspective can be an incredibly useful tool to help us better understand contemplative nutrition, as such a philosophy puts the environment and the needs of others at the forefront of our food-related beliefs because ultimately what affects one affects us all. Like it or not, we live in an increasingly interconnected world where climate change and other environmental concerns are planet-wide problems. If you were raised in an individualistic culture, learn to acknowledge this and be mindful that this will bias how you view your food preferences. Whenever you experience food-related anxieties, pause and ask yourself, 'Am I being me-focused?' This will help you to make better choices. Consider how others and the environment might be affected by your food-related decisions. This is contemplative eating.

In the 21st century, despite political instability, the looming climate crisis and geopolitical tensions, the vast majority of people have complete food security. When it comes to having enough nutritious food available, most of us have less reason to be anxious than at any time in the past. Recognising this fortuitousness enables us to be mindful of our anxiety bias and to notice that we need not be suscep-tible to our insecurities, in the knowledge that we will enjoy our meal and, even if it doesn't quite live up to what we hoped, we can appre-ciate something else just a few hours later. Instead, we're confronted with other anxieties: the impact of the food system on the environ-ment, our health and our wellbeing. Modern agricultural practices and unfavourable food choices present real and grave concerns. Writing in *The Economist* in 2021, and referring to the current geolog-ical era where human activity is the dominant environmental influence, journalist Jon Fasman describes 'the Anthropocene diet'. Fasman notes that what the world's well-off inhabitants enjoy today would have amazed all previous generations. He looks to technolog-ical advances, like those presented in the last chapter, to help deliver cleaner, greener and more delicious food. He is, however, unsure if consumers will want it. Fasman writes, 'Deciding first what sort of person you want to be, and what sort of planetary settlement you want to embody, and then changing the world so that the kind of food it provides for you to eat fits that self-conception'.[4] In essence, we need to think about what we eat and the effects this will have. This is exactly what contemplative nutrition is. All we need to do is cut down on animal products, buy better and avoid eating crap, right? Oh, if only things were this straightforward …

The Dark Side of Sustainable Eating

Following several visits to the doctor, multiple syringefuls of blood and numerous pokes and prods, there was increasing concern as to why Hannah had been feeling weak, struggling to concentrate and experiencing weird wrist and ankle pains. Her teachers had reported that her work hadn't been up to her usual high standard, and, at home, her parents had put her mood swings down to 'just being a teenager'. Worryingly, some investigations had been for neurological impairment, prompted by a family history of motor neurone disease. While waiting for the results of medical investigations, there had been considerable turmoil at home.

As it turns out, the reason for her symptoms was poor nutrition. Over the past several months, Hannah had become increasingly concerned about issues like climate change compounded by the GHGs emitted from our food system. Eager to change her lifestyle and to 'eat for the planet' – and inspired by Swedish activist Greta Thunberg – she'd adopted a plant-based diet, but with little concern for the adequacy of her own nutrition. When I heard this torrid tale from Hannah's family, parallels with the 'eating clean' phenomenon immediately sprung to mind.

The Problem with 'Clean Eating'

We all recognise the importance of consuming a nutritious diet in order to keep our minds and bodies in shape, but when taken to the extremes, things can take an unfortunate turn. Living in a world where the infectious allure of high-profile figures overly hype the perceived healthfulness of particular foods, supplements and dietary regimens, one could be forgiven for believing that religiously strict 'clean eating' protocols are essential if you want to achieve optimal performance. Trendy diets are often highlighted on social media, typically by non-expert celebrities, yet there's no scientific consensus around what constitutes 'clean eating'. A 2020 study found that people often associate 'clean eating' with extreme weight-loss diets, with horrific consequences on their health and mental wellbeing.[5] Definitions of 'clean eating' are varied and typically include elements such as consuming local, organic, plant-based, home-cooked or minimally processed foods, but frequently also encompass more extreme strategies, like eliminating gluten, grains, carbs or dairy.

Disentangling the myths and horrors around what matches the definition of 'clean eating' is complex and the whirlwind of misinformation flogged to us each day belies opportunistic behaviours in the food industry, sometimes with grave consequences for our health.

Skimming the cream off the top rather than diving into the scientific literature has obvious advantages: few of us have the time, inclination or necessary skills to indulge in hundreds of highly technical reports. But this approach bestows a bigger problem: faced with an overwhelming amount of bite-sized, Twitter-friendly information, many are no longer able to distinguish between healthy food and fads. We live in the age of online influencers and all ideas – however steeped in science – have to battle glamorous social media stars. Commenting on clean eating, Anthony Warner, a food writer and scientist by training, once said that this type of diet 'tends to be a hotbed of strange pseudoscience, used to justify exclusion driven diet regimes'.[6] History seems to prove him right. A tragic case in point comes from Australia. In 2008, food magazine journalist Jessica Ainscough was diagnosed with a rare form of cancer called epithelioid sarcoma. Disenchanted with the efficacy of conventional medical treatments, Jessica opted for a rather unusual treatment called 'Gerson therapy', a strategy developed in the 1920s.[7] This treatment involves following a strict organic, vegetarian diet and numerous dietary supplements and daily coffee enemas. Academics around the world have repeatedly rejected Gerson therapy with the website of Cancer Research UK stating: 'There is no scientific evidence that it can treat cancer or its symptoms'.[8] Tragically, Jessica succumbed to her disease in 2015. Experts agree that with medical treatment she would likely still be alive today. More recently, in August 2023, fruitarian social media influencer Zhanna Samsonova died of starvation after following a diet where she ate only raw fruits, supplemented with a few vegetables and seeds.[9] Although Zhanna and Jessica provide extreme examples they demonstrate the importance of heeding advice from qualified professionals and snubbing ultracrepidarians.

Similar issues abound, with growing evidence showing that so-called 'healthy eating' may disguise an infatuation for the purity of food, causing problems in work, social and emotional functioning. This obsession is known as 'orthorexia' and is characterised by compulsive behaviour, self-punishment and what some scientists dub 'right' food addiction.[10] Though yet to be formally recognised as an eating disorder, orthorexia bears many similarities to commonly known disturbances in eating behaviour like anorexia. Circumstantial evidence suggests that this condition is common among ballet dancers, athletes and health professionals.[11] The desire to gain a more athletic body contributes to inflexible eating habits. Although no single dataset exists around the prevalence of this condition, available studies point to staggeringly high estimates across countries, ranging from nearly 7 per cent in Italy to 89 per cent among Brazilian students.[12] While its name is still missing from clinical textbooks, consensus is growing that orthorexia may lead to complications similar to those of formally recognised eating conditions. Research shows that a hyper-focus on perceived healthy eating can lead to malnourishment as orthorexic individuals tend to engage in dangerously restrictive eating practices, often excluding key nutrients from their meals.[13] More worrying still, someone with orthorexia may withdraw from their social life and meaningful relationships in their pursuit to eat 'healthy' and 'clean', paving the way to social and psychological problems. Orthorexic eating behaviours may further elicit forms of anxiety, depression and guilt, especially in the wake of perceived 'violations' of their strict eating practices. One study pointed out that up to 48 per cent of individuals with orthorexia suffer from at least 'moderate depression', with one in three showing obsessive-compulsive symptoms.[14]

Why is this important? As is the case for the gut–brain connection, academic discussions of orthorexia and clean eating are littered with references to huge bodies of research that still need to be done. What we know suggests that the line between healthy eating and developing an eating disorder is worryingly thin, with orthorexia often starting as a result of an innocent desire to improve one's food habits or to eat more sustainably. This means that the obsession with health is not confined to a few fanatics: everyone can become obsessed to a certain extent, and our fad-crazed food industry really isn't helping. Over the past decade, there's been a shift in food marketing towards organic and natural products, and consumers in developed countries are increasingly demanding minimally processed food, in the fervent belief that such food is healthier for their bodies. This major shift in behaviour is rooted in the dominant ideology within contemporary wealthier societies that health is a moral imperative. Time-strapped employees and employers have rediscovered the mantra, originally made by the Roman poet Juvenal, *mens sana in corpore sano* – 'a healthy mind in a healthy body' – fuelling the modern craze for fitness, health and everything in between.

Non-Sustainable Eating

In the 21st century, sustainable nutrition and healthy eating are intrinsically tied. To consume a good diet means considering your own health, the wellbeing of others and the environment. Eating sustainably involves making food choices that have minimal impact on animals and the environment in order to protect biodiversity, ecosystems and communities, both now and for future generations. We've also seen that diets that harm the environment are also often of poorer nutritional quality.

Crucially, a sustainable diet must support your own health. Yet looking after your own health is more than just taking care of your physical body: cognitive and emotional health are equally important. If we truly care about others, a sensible strategy must primarily consider our own wellbeing. Prioritising ourselves is imperative: if our own health is subpar, how can we effectively look after others? If eating a nutritious diet helps to keep us out of hospital, we free up resources to be better spent elsewhere. By definition, a 'sustainable diet' isn't sustainable if it can't sustain an individual.

Contemplative Dilemmas

Eat carbs; avoid carbs. High protein is good for you; protein damages your kidneys. Use margarine; have butter. Eat plenty of fruit; fruit's full of sugar. Avoid overeating; don't throw food away. Consume more oily fish; overfishing is a problem. Have meatless burgers instead of beef burgers; avoid processed food. Eggs are good for you; eggs raise cholesterol. Coffee is full of good stuff; cut down on coffee. Red wine is rich in antioxidants; cut down on alcohol. Eat plant-based; go carnivore. I could go on …

Confused? Oh, if I had a penny for each time I heard someone say, 'Nutritionists are always changing their minds!' As we saw earlier, nutrition is a minefield of misinformation, and things get even more complicated when we add 'eat sustainably' into the mix. While much of the contradictory advice we hear stems from pseudoscience, hidden agendas and dietological incentives, other controversies can be more rational. Science is progressive and this means dietary advice will change the more we learn, and this will lead to constructive disa-greement. No one knows what an 'ideal diet' is or even if there is such a thing. But it's the same in every facet of life: each day, conflicting pressures give rise to numerous zero-sum dilemmas. Do I knock off work at 5pm to get home in time to read the kids a bedtime story, or do I stay late to finish the report to keep my boss happy? Do I save an extra £100 this month knowing that the interest it earns will compound, or do I buy those jeans? We're confronted with similar dilemmas when we're deciding what to eat. That scrumptious slice of cake now versus a tiny step towards my weight-loss target?

Contemplative nutrition involves considering the best options based on multiple perspectives. There are no solutions, only trade-offs. Consumption of certain foods may be in line with one aspect of contemplative nutrition, but might conflict with another. What do we do about foods that have a possible health benefit, but are unsustainably sourced? What about a certain dietary strategy that might boost performance, but adversely affect our mood? Let's look at a few of these potential conflicts and see how we can find a workaround. There will be many more not covered here but by taking a reason-based perspective when it comes to food choice you'll be able to navigate any potential dilemma.

Waist or Waste?

During the early 1990s when studying to be a dietitian, my training included advising people who needed to lose weight. Guidance involved helping clients to recognise when they felt sufficiently full and, at that point, to stop eating. Strategies included serving food away from the table to avoid the urge to indulge in second helpings and to limit the amount they dished up onto their plate in the first place. Clients were advised not to feel obliged to eat everything and to throw leftovers in the bin right away so as to avoid the temptation to finish them. These sensible tips aimed to limit how much they consumed to help them control their weight. Many of the people I advised at the time were from the generation that had been raised in the era of post-war rationing when food was often hard to come by. Advising them to chuck food in the bin went against their values.

Now, nearly 30 years on, discarding food is once again a no-no but this time for different reasons. As well as its ecological impacts, food waste has been directly linked with calorie excess. In a 2019 paper, the authors claimed that 'consumption of food that is constantly above the recommended calorie requirements by a growing number of people not only represents a risk to health, but it also puts more pressure on natural resources and on the environment'.[15] Using the Metabolic Food Waste index, which calculates the ecological impact of obesity per person,[16] they demonstrated that increased food intake is accompanied by a disproportionate increase in food waste, meaning that obesity contributes to climate change as a result of unnecessary food consumption. Obesity should, they assert, be recognised, not just as a health and social issue, but as an environmental problem too.[17]

Binning food does seem counterintuitive: after all, our ancestors struggled to acquire sufficient sustenance for millions of years. Should, then, these calories end up on our waist or in the waste? The answer seems obvious: plan ahead and don't cook more than you need. This way the amount of food you waste will be minimal. It's not so straightforward, however. The modern mindset favours preparing more food than you need. Recently, when wondering how much food would be required for a party, a friend stated, 'It's always better to prepare too much than too little!' While I applaud her eagerness to fulfil her guest's gluttony, it's this kind of societal mentality that needs to be thrown away. Maybe it's a mindset that stems from an innate urge to always have some food in reserve.

What about when there's food in your fridge that's about to go off? Planning meals ahead, not overbuying and using up the short-dated produce first will help to minimise what's wasted, and veg that's about to go off can be used in soups and stews. But what if you're limiting the amount of meat you consume? Let's say you've got a steak left in the fridge that will go bad if it's not eaten today. But what if you've already met your weekly 10 per cent animal calorie allocation and gorging on another sirloin would take you well over? The sensible thing to do is to eat it. Why create unnecessary waste? And binning it would only further devalue the cow's sacrifice, not to mention have been a waste of your hard-earned cash. As a countermeasure, buy less meat next week.

What about when a friend prepares a meal that contains ingredients you choose to avoid? Say, without knowing, she serves a dish containing meat that's not from a free-range animal. How you navigate such a scenario is ultimately your decision but, unless you're a strict vegetarian or vegan, the rational thing to do is to eat the meal or, at least, some of it. By refusing it, not only do you risk offending your host, but you're also contributing to unnecessary food waste. Make a mental note to mention your restrictions in a future conversation.

Go Teetotal?

A controversial 2018 review published in *The Lancet*, definitively titled 'No Level of Alcohol Consumption Improves Health', concluded that any potential beneficial effects of alcohol were more than cancelled out by the negative effects.[18] The paper was criticised primarily for two reasons. Firstly, the title implied a fact and epidemiological studies should be observational, i.e. show correlation rather than causation. And secondly, well, who wants to hear such devastating news?[19]

Since the so-called 'French Paradox' of the 1980s, which led to the belief that drinking red wine was responsible for the lower incidence of cardiovascular disease in the French despite their diet typically being high in saturated fat, many people, researchers included, adopted the belief that drinking a moderate amount of alcohol can be 'good for you'.[20] We've seen how alcohol affects cognition and, as far as health and longevity is concerned, the consumption of *any* amount of alcoholic drinks has been linked to an increased NCD risk. Alcohol is, after all, toxic and technically a poison. The reason why being French in the 1980s might have been good for your heart could be any one of a multitude of reasons, such as – among other hypotheses – the benefits of a Mediterranean-style diet, more sunshine and more vitamin D, or a more relaxed and less stressful lifestyle.

The benefits of community and socialising, especially around food, are probably as least as beneficial to our health as good nutrition. If having a little alcohol helps you to feel a little less anxious, allowing you to feel more at ease around others, then maybe it's okay to have the odd drink. Drinking is so ingrained as a social norm that a total abstinence from alcohol could risk feelings of exclusion. When it comes to deciding whether or not to drink, contemplative nutrition means being aware of the risks and benefits and making an informed decision. It might also be wise to acquaint yourself with an unbiased perspective on alcohol.

Nutrition Science Isn't Rocket Science

Although we've seen that processing food can have huge benefits, we know that a large number of the commercially produced junk foods are not best for our health. Trans fats, for instance, now unanimously acknowledged to be a health hazard, just a few decades ago were touted as safe. With this in mind, how can we be certain that some of the constituents of our favourite foods won't in the future be revealed to be dangerous? Thankfully, these days, novel ingredients are subjected to intense scrutiny before being permitted in our food. (Indeed, these stringent regulations add frustratingly long delays to the release of many exciting and much required innovations.) Enjoying the occasional chocolate bar, pizza or cookie won't cause you any long-term troubles, of course, and rarely, if ever, should a food be demonised. But is there anything we should be watching out for? 'Trans fat' or 'hydrogenated oil' on food labels, of course, and, fortunately, due to both legislation and consumer pressure, they're now rarely encountered. Another indicator of less nutritionally desirable products are those where 'sugar' appears high in the ingredient list.

Where possible, it can also be useful to know the preparation methods involved as these can affect the chemistry and the nutritional value of the ingredients. An example is extra-virgin olive oil, with its light, subtle taste and monounsaturated-richness. Virgin oils are delicate and easily damaged by heat or light. Frying can cause them to oxidise, ruining their flavour and increasing their atherosclerotic potential. Similarly, some polyunsaturate-rich seed oils, such as flaxseed oil, when not stored correctly can become damaged, throwing their otherwise omega-3-goodness out the window.

How can you tell if the food on which you're about to gorge will be okay? How can a layperson find out if it contains bad fats, refined carbs or ingredients damaged by the processing? Simply put, often we're forced to resort to good old common sense. Nutrition science isn't rocket science and more than likely you've already got a pretty good idea which foods you might be wise consuming less frequently and only in moderation, especially now you know to watch out for misinformation-spreading pseudo-influencers. Those mouth-watering junk fast-food snacks might tantalise your taste buds, but gorging on them too often might terrorise your arteries.

Not Enough Fish in the Sea

We've seen the benefits of omega-3s in relation to cardiovascular health and disease risk, and we've seen that the fatty acids DHA and EPA, predominantly found in seafood, may help cognition and focus. We should, therefore, favour recommendations when they advise us to include oily fish – such as salmon, sardines, pilchards, mackerel and trout – two or three times in our weekly diets. However, with fish populations under threat from the effects of global warming and overfishing, and concerns that aquaculture might not be the fix-all advocates claim it to be, we're presented with a quandary. We're rightly told that eating oily fish is good for us and, as previously stated, our own health should be prioritised over other considerations, yet the practices used to obtain these fish are neither justifiable nor practical. Welfare issues aside, to put it bluntly, there's not enough fish in the sea. Currently six out of ten of the UK's most important fish stocks are overfished or in a critical situation.[21] In 2019, on average, Brits each consumed more than 150 g of seafood (around a portion) a week.[22] The UK recommendation is at least two portions of fish a week, one of which should be oily fish.[23] This is the bare minimum so, from a health perspective, we're already not consuming enough. Despite the recommendations being updated in 2016 to suggest that the two portions should come from sustainable sources like MSC certified fish,[24] this only addresses the sustainability concerns, saying nothing about fish welfare. Moreover, it's not clear that sustainable sources of seafood are always easily available. So, it's not just the herrings that are in a bit of a pickle!

With insufficient numbers of fish available to meet the already inadequate health recommendations, how do we intend to acquire precious omega-3s? Sure, we've established that molluscs, like mussels and scallops, are sustainable alternatives, but these contain only modest amounts of omega-3s. Once again, we look to plants for help. Essential ALA and other beneficial omega-3 fatty acids are present in numerous plant foods. Granted, many of the crops available in today's supermarkets contain rather unimpressive amounts, but reasonable quantities can be obtained from certain nuts and seeds, like walnuts, hemp seeds, chia and flaxseed. We can also look to where the fish themselves acquire their supply of omega-3: algae. Consuming edible seaweed or algae supplements can supply ample DHA and EPA at an affordable price. Algae can be produced in bioreactors or can be farmed in the seas, where it has another cool attribute: it's really good at sequestering carbon.

Less Carbon or More Suffering?

We've illustrated the deplorable practices of factory farming and presented a case for humane alternatives. However, when it comes to GHG emissions and livestock farming the debate is less clear. From a global perspective, intensive farming emits fewer CO_2e than grazing animals. The food with the widest GHG disparity is beef, so it's a useful example. The worst beef emits around 105 kg CO_2e per 100 g protein, the best only 9 kg.[25] Although other foods have a much narrower spectrum, particularly plant foods, the best beef still fares worse than the worst plants. Buying locally is often seen as an attempt to minimise emissions because many people believe that most of the CO_2e emissions are the result of transport and packaging. In reality, these account for less than 2 per cent of beef's total emissions[26] and nearly all food transport emissions come from the last few miles.[27] So although buying your free-range meat locally might seem like a good idea, from a carbon perspective it might not be the best. As we've seen, methane produced from cattle is a huge contributor to climate gases, but it makes a difference where the beef comes from. For instance, beef that originated from a dairy herd contributes less methane than from a herd dedicated to beef production. Forty-four per cent of beef comes from dairy herds, sharing its environmental impact per calorie and per gram of protein with milk. Moreover, dairy cows tend to get higher-quality feed making them grow faster and emit less methane.[28] Also important is the type of land on which the livestock are farmed. The worst by far is the destruction of forests for farmlands as not only does this release the CO_2 that was bound in the flora, it also sets soil-stored carbon free and ruins its ability to store it in the future. Additionally, fewer trees means less CO_2 can be sequestered from the atmosphere.

It seems that the greater the degree of animal suffering, the lower the impact on climate change. How do we navigate this sinister conundrum? More efficient systems of farming means less land is used and fuel efficiency improves. Add to this the fact that cows confined to factory farms never get to roam pastures and so are less destructive than cattle grazing peacefully on a formerly lush piece of rainforest. However, half of all agriculturally used land is dedicated to animals, most of which is grassland that can't be converted to cropland,[29] so grazing cattle on these pastures allows them to turn grass, something we humans can't digest, into food that we can. From an economic perspective, farming animals can seem like a smart way to make the best use of unused resources. Globally, grazing systems sustain only 13 per cent of beef production,[30] and although pastures could support more livestock than they do, this isn't necessarily desirable and the world's grasslands are not infinite. If global beef consumption were to switch to 100 per cent grass-fed cows, we'd have to eat much less beef. It's a similar story for other livestock like pigs and chicken. Even if you found the most sustainability sourced beef in the world, your burger would still come with a significantly higher carbon tag than a meatless patty. The wise strategy is to cut right down on the amount of meat you eat, sticking to meat from livestock reared on pasture where there's been no forest destruction. Better still: opt for plant-based alternatives at least until lab-grown meats are affordable and available in stores.

Lower Carbon or More Deaths?

Cattle farming is the largest contributor of GHG emissions. Beef produces more CO_2e than any other livestock per kilogram with chicken being the most favourable when it comes to GHGs.[31] From a carbon perspective, it makes sense to opt for chicken over beef. But what if you care about sacrificing the fewest number of sentient beings possible? By consuming chicken, more lives are lost – more than 150 chickens for every cow.* Is it more ethical to stick to beef, sacrificing one cow to 'save' hundreds of chicken lives, or favour chicken and commit less GHG to the atmosphere? There is no correct answer here. Maybe the solution is to minimise your consumption of both.

How Much Animal Produce?

We've seen that prioritising nutrition from plant sources has significant environmental benefits. But we've also seen humans have an innate desire for meat. It's understandable, therefore, that many people will be reluctant to quell their steak eating altogether. Reducetarianism, a term coined by Brian Kateman, founder of The Reducetarian Foundation, is the process of reducing meat, dairy and egg consumption without excluding them completely. The organisation offers support to those committed to reducing their consumption of animal produce and 'aims to improve human health, protect the environment, and spare farm animals from cruelty by reducing societal consumption of animal products'.[32] Like flexitarianism, reducetarianism is aligned with my suggestion to limit animal products to 10 per cent of energy intake. Many argue, however, that this reduction is insufficient, correctly stating that even though the reduction in environmental impact is significant, there will still be carbon, land and water consequences. Moreover, reducetarianism still permits what many see as the exploitation of animals for human desires, albeit to a lesser degree, even when actions include the boycotting of

* This estimate is calculated from approximately 1.5 kg chicken meat per bird vs 225 kg beef per cow, and assuming a similar portion size.

factory farm produce. Indeed, these objections are even more valid when we acknowledge that humans are adaptivores: as a species with the ability to acquire nutrition from a wide range of foods, humans can thrive without the need for animals.

Reducetarian, flexitarian, vegetarian, pescatarian and 100 per cent plant-based strategies can all be examples of contemplative diets if they're based on a reasoned consideration of nutrition for health with minimal impact on others and the environment. But what about veganism? Can a vegan diet be contemplative nutrition? After all, going vegan mitigates many of the environmental and ethical hurdles. But what about the effects of veganism on our own health and well-being?

To Vegan or Not to Vegan? That Is the Question

Is it acceptable for 21st-century humans to continue to eat meat or should we be completely expelling animal products from our plates? Of all the dietary dilemmas, this is possibly the most controversial. Michael Pollan spent more than 400 pages debating the issue in *The Omnivore's Dilemma*[33] and the internet is brimming with vegan versus meat-lover rivalry. Meatless diets, interestingly, aren't something that animal activists have recently conjured up. Pythagorean diets, as they were originally known, have been around since at least 500 BCE, although the term 'vegetarian' didn't show up until the mid-1800s and 'vegan' only made its debut in 1944.[34]

Our food choices are associated with our identity. Massimo Montanari describes food as a lexicon for a culture's tastes and values,[35] meaning that what we eat is an expression of our own identity in relation to our beliefs. We attribute value to what we eat, as illustrated by the importance of certain food practices in religions. Veganism is an example of a value-based belief system. Vegans adhere to a way of living that excludes all products – food or otherwise – derived wholly or partly from animals. Adherents to veganism believe that animals should not be exploited in any way, and this includes not consuming their flesh or anything produced from or by any creature, as well as refraining from using any product where an animal has been exploited in any way, such as clothing made from leather or wool or cosmetics that have involved animal testing. This relates to all sentient beings and mandates the non-consumption of all meat, fish, insects, milk and dairy products, eggs, honey or any product containing ingredients derived from the above, such as gelatin. The Vegan Society define veganism as

> *a philosophy and way of living which seeks to exclude – as far as is possible and practicable – all forms of exploitation of, and cruelty to, animals for food, clothing or any other purpose; and by extension, promotes the development and use of animal-free alternatives for the benefit of animals, humans and the environment. In dietary terms, it denotes the practice of dispensing with all products derived wholly or partly from animals.*[36]

A plant-based diet is one that consists almost entirely of foods derived from plants, and may include artificially derived ingredients, such as those involved in food processing or added vitamins, or even tiny quantities of animal-derived ingredients. Plant-based eating can also be a value-based belief system, if the decision to eat plant-based considers certain values, such as concerns relating to environmental destruction. We've yet to conclude if a vegan diet is an optimal dietary strategy when making considerations relating to health, wellbeing, the environment and ethics simultaneously, or, indeed, if 21st-century Homo sapiens *should* be vegans.

To understand if humans should be vegans, let's begin by looking at a defining feature of our species. Humans are social foragers: we share the foods we hunt and gather. Early during the transition to the genus Homo, when we started to use stone tools, meat became a regular menu item. Vegetarian species typically don't share their foods because foraged plant foods are less energy-dense than animal acquisitions making it costly for them to be shared. Sure, chimps and bonobos do on occasion share with their peers, but it's rare, with reluctance and generally only when the rewards of hunts include monkeys, small antelopes or the occasional large fruit. However, for hunter-gatherer humans, sharing has been ubiquitous and advantageous, demonstrated by tubers which, although troublesome to forage, supply large amounts of energy implying that it's useful for them to be shared. It's the combination of hunting and gathering that has helped humans become the altruistic sharing species that we are.

Acknowledging that our hunting-gathering-combo ancestry has contributed to our benevolence might lead to the implication that humans are 'meant' to be omnivorous. Yet we should examine this claim in more detail. Humans are adaptivores, and our adaptations have resulted in part from our genetic and cultural evolution, and this has forced meat into nutritional significance. While scouring the land for plant foods meant merely foraging close to home, hunting for flesh often mandated venturing further afield. These demands led to numerous anatomical and metabolic adaptations. Successfully running and chasing down prey over long distances involves super-efficient conversion of fuel into energy. Our dentition and alimentary canals have evolved to allow us to break down animal flesh efficiently. By being endowed with teeth designed to tear meat from bone and then to grind it down, and with a relatively short gut with an enlarged intestinal surface area, we're assured of a highly efficient uptake of nutrients. Our sensitive tongues have receptors tuned so acutely that when they're stimulated by meatiness, messages are sent to our brains that make us experience pleasure. We crave the familiarity of the succulent aroma and luscious flavour of cooked meat. Meat is a form of sustenance actively sought out by humans. As well as this, the vast majority of human cultures – current and historical – have enjoyed meat as a key part of their social habits. The communal consumption of meat has aided social bonding, cemented friendships and helped to build bridges with former foes. Even today, eating meat is seen as 'normal' behaviour. The authors of a 2021 paper observed that people's national social identification was a significant predictor for meat-eating: 'Americans do love their hotdogs, the British do love their Sunday roasts, and Australians love their meat pies. Levels of meat consumption in these countries are objectively high and … allowed for the perception of a high descriptive norm for meat eating'.[37] When people grow up with meat as a staple, when meat is available in stores and restaurants, when hospitable hosts serve meat to their ravenous salivating guests and when we watch those around us take enormous pleasure from devouring meat the presence of meat as a food becomes increasingly normalised.

Taking into consideration that we have brains that tell us to seek out meat, physiologies designed to process meat and societies whose advancement has been dependent on meat, surely it must be the case that humans *should* eat meat? On the other hand, we've demonstrated that, as adaptivores, humans can readily obtain more than sufficient calories and adequate nutrients from a large variety of foods without needing to consume any animal products. Indeed, such diets can be similarly pleasurable. Moreover, the fact that cultural eating habits have evolved tells us that adaptations are fluid and cultures could continue to evolve without the inclusion of meat. Even an emotionally driven decision to become vegan could be considered contemplative nutrition if the thought process was based on objective reflections. Peter Singer, in his 2007 essay *The Case for Going Vegan*, argues that veganism is actually a rational choice: 'If there is no serious ethical objection to killing animals, as long as they have had good lives, then being selective about the animal products you eat could provide an ethically defensible diet.' Singer noted that taking on a vegan diet needs care, however, and that the term 'organic', for instance, says 'little about animal welfare and hens not kept in cages may still be crowded into a large shed'. 'Going vegan', according to Singer, 'is a simpler choice that sets a clearcut example for others to follow'.[38]

Thus far, we've failed to answer the question *should 21st-century humans continue to eat animal products?* Could it be that we're attempting to tackle the question in the wrong way? Maybe we shouldn't be looking to either biology or anthropology for an answer. After all, biology is telling us that we should consume animals while at the same time declaring that we don't need to. Similarly, anthropology informs us that as our ancestors relied on meat, we should continue to eat it, yet simultaneously hints that we no longer need to. Instead, maybe we should look to morality to provide the answer. Yet, the moral argument, while somewhat compelling, is still non-committal. We have depended on meat in the past, and we should, indeed, be grateful when we acknowledge how pivotal it's been to human flourishing. As we no longer need to consume meat, and plant-based diets are far less environmentally destructive and minimise harm to nonhumans, any moral analysis would compel us to seriously consider if we should continue to eat animals. Moreover, even when we invoke moral arguments in favour of veganism, urging people to give up meat is going to be met with considerable pushback, especially as it seems morality is also failing to provide a definite answer to our question. If according to the morality argument eating nonhuman species is immoral, why has it been paramount to human flourishing thus far?

My conclusion, therefore, is disappointing. I don't believe we are able to answer the question *should 21st-century humans continue to eat animal products?* At least for the time being. Maybe the day will come when we can claim that eating animals is morally wrong. At some point, we may be able to confidently assert that the 'right' diet is one that avoids any kind of animal exploitation. I'm confident, however, that, in order to reach this point, we will be dependent on food technological innovations. For now, instead maybe we should be content to acknowledge that including some animal foods in our diet is morally 'acceptable'.

My justification for failing to provide an answer to this important question lies in contemplative nutrition: choosing what to eat should include contemplating your position on the consumption of animal products. While humanity catches up, however, we can cut down on the amount of animal produce we consume. A plant-rich diet that includes small amounts of animal products is a rational solution. After all, let's be honest, the principal reason why most 21st-century Westerners voraciously gorge on copious quantities of meat is gluttony, and to feed this meat-greed we are reliant on exploitative industrial farming with all its inherent cruelty. Through invoking a moral perspective, we should be transitioning to only consuming meat that's produced from non-industrial sources. Pasture-reared animals enjoy a better quality of life, can have a lower environmental impact, and their meat is more nutritious, and, some reckon, tastier, too. Seeing as we can't say for certain if we 'should' or 'shouldn't' be consuming meat, reducing how much we consume and sticking to meat produced by more humane practices seems like the rational and sensible compromise. For now.

Not Contemplating a Contemplative Diet

Although since my late teens I've paid attention to my nutrition, only recently have I begun to consider more deeply the wider implications of what I eat. While prioritising health, wellbeing and performance goals, I see no need for food choices to cause unnecessary excessive environmental effects or undue suffering. I enjoy food: the flavour, texture, aroma, appearance and convenience of food all have a bearing on what I choose to consume. But these are secondary to nutrition, environmental and animal welfare considerations. Contemplative nutrition has evolved out of my own way of eating.

Now that we've explored the science-led and philosophical influences on food choice, we need a practical set of guidelines of how to eat better. Few of us will be enthralled by the idea of having to scrutinise everything we propose to put into our mouths. Most of us are already paying attention to what we eat at least to some degree: we recognise convenience, nutritional value and the pleasure we glean from food. Contemplative eating needn't be hard: we just need to be steered by a few basic rules that support optimal physical health, mental wellbeing and performance, whilst addressing the concerns regarding animal welfare and the environment. If always having to think carefully about what you eat sounds like a lot of bother, that's because it is. Our ancestors simply ate what was available, sparing little thought other than to how they would acquire their next meal. Yet, our food system is far from perfect, so if we value our health and humanity's future, we should at least be giving *some* consideration to our food choices.

There are six key focus areas of contemplative nutrition:

1) Choose nutritious food for long-term health

Good-quality, nutritious food is essential for your physical and mental health, and what you eat will have a bearing on your physical performance. Good nutrition needn't be overly complicated: most of us already know what we should and shouldn't be eating. Although some official nutrition recommendations could be challenged, broadly speaking, the principles of healthy eating policy are sound, and following them will ensure a good nutritional intake. Choose a varied diet with a range of different foods, including plenty of fruit and veg, pulses, nuts and seeds, wholewheat cereals and grains, and potatoes, as well as, possibly, eggs and a limited intake of lean meat or fish. Junk-food meals (we all know what we mean here) should be kept to a minimum and junk snacks should be kept as treats. Despite what some fearmongers say, there are no foods you must have and few that you shouldn't have.

. If you don't possess them already, learning basic cookery skills will be hugely helpful. Purchase in bulk and use a freezer to help with both cost and convenience; frozen fruit and veg are great. If you're physically active, adapt what you eat to suit your goals. This needn't be too regimented: simply eat good-quality food and adjust appropriately, with a focus on protein-rich and slow-released carb foods, with plenty of foods rich in fibre and essential fats. Adjusting *when* you eat may also help you to optimise your performance. There is no one-size-fits-all advice for good health. As much as people want conveniently laid-out meal plans, we're all biologically different and have unique lifestyles and goals.

2) Structure your diet to maximise your wellbeing

What and when to eat varies between individuals. For those with busy lifestyles, food convenience is important. Gym-goers, focused on improving performance or making gains, may benefit from a particular structure to their diet. Others may wish to adopt strategies such as intermittent fasting. Alcohol is best limited to social drinking and hot drinks can be enjoyed as long as caffeine doesn't affect your sleep quality. While it may not always be possible to have meals with others, where circumstances allow, try to eat with friends or family two or three times a week. Cooking a meal from scratch and eating it together is fun. Preparing food for others is a fundamental act of kindness, and what a special privilege it is to be invited into someone's home and for them to serve you food. Eating out is great too and can be a valuable part of our diet. Social eating, in varying ways, has been part of every human culture for millennia and is often based around nutritious food relevant to the region in which the culture originated. Take advantage of all those lovely cuisines from different parts of the world.

3) Limit consumption of animal products to around 10 per cent of total energy

In the preceding chapters, I've provided a rationale for limiting the amount of meat, fish, dairy and eggs we consume. As adaptivores, we don't need to consume anywhere near the amount of animal-derived ingredients that we do. Reducing animal products to 10 per cent of total calories could give as much as a 35 to 45 per cent reduction in your food-derived carbon footprint. This provides ample animal-derived nutrition, enhances variety and the range of flavours adds to the enjoyment. And if you feel a more plant-based diet is for you, go for it.

4) Avoid intensively farmed animal products where possible

If you object to the conditions in which intensively farmed animals are kept, avoid products from animals reared by such methods. There are numerous butchers who sell pasture-reared meat, both locally and online, and farmyard-roaming chicken's eggs are plentiful. The meat may be a little more expensive, but it's more nutritious and, as you'll be limiting the amount of meat you're eating, the cost impact for a meal will be minimal. If you're unsure about the conditions in which animals have been farmed, then it's best to avoid the meat. If you include seafood, where possible favour molluscs and sustainably sourced fish.

5) Don't waste food

Around a third of food produced for human consumption is wasted every year.[39] We can't control what others do, nor can we do much to influence food manufacturers, stores and restaurants, but we can limit the amount of food we waste at home. Western society has this ridiculous notion that it's better to buy too much than not enough. This mindset encourages wastefulness. Don't overbuy: plan ahead and consider shopping locally. This may mean two or three small shopping trips each week, but you'll throw less away and save money. Also, buy in bulk and freeze. There's lots you can do to minimise waste. Veg nearing its use-by date can be used in stews and soups, and you'll be surprised how little of an animal's carcass you need to waste. For example, once you've stripped most of the meat from a roast chicken, use the remnants to make bone broth with vegetables, pulses and potatoes; bone marrow is very tasty and rich in valuable nutrients.

6) Enjoy eating!

Although the focus has been to prioritise science-based and ethical judgements over organoleptic considerations, we must, of course, enjoy our food. Humans have evolved to derive pleasure from different flavours and textures. Likes and dislikes vary and contemplative eating involves socialising around and enjoying nutritious food. Take pleasure from your meals.

Contemplative Nutrition

The following is an amalgamation of the key call-outs from the chapters arranged into 20 contemplative nutrition tips:

- Make choices that prioritise your own health.
- Plan meals ahead and don't buy more than you need to.
- Buy in bulk; utilise a freezer.
- Eat a wide variety of foods.
- Have regular meals and try not to overeat.

Contemplating What to Eat

- Base meals on tubers, wholegrains, pulses, nuts, seeds, fruit and vegetables.
- Include fibre-rich foods.
- Consume foods rich in omega-3s.
- Limit animal-derived calories to 10 per cent by moderating your meat and seafood portions and including two or three days a week where you almost entirely stick to plants.
- Avoid meat, poultry and eggs from factory farms, favouring produce from game or animals reared more humanely.
- Seafood may be included: opt for molluscs and sustainably sourced fish.
- Limit dairy: choose plant-based milks or organic milk and butter from grass-fed cows.
- Embrace food technologies while watching out for junk processed foods with added sugar and hydrogenated fats.
- Where applicable, favour products certified with ethical credentials.
- Try not to waste any food; again, plan ahead.
- Fasting for prolonged periods may be worth trying to see how you get on.
- Hot drinks have feel-good effects.
- Enjoy alcohol in moderation and only occasionally.
- Heed the watch-outs for quack influencers and approach food marketing claims with caution.
- Socialise: whenever possible enjoy meals with others, sharing the meal preparation.

A 'Perfect' Food System ... Revisited

I started this book by refuting the claim that the human food system is 'broken'. I acknowledged that it's far from perfect, but noted that never in history has it been 'fixed'. I pointed out that during recent decades the global food system has improved and, for the first time in history, everyone could have access to sufficient calories and protein from a varied range of foods. I showed how the food system is moving in the wrong direction, illustrated by its role in environmental destruction, questionable livestock practices and metabolic diseases of abundance. I also attempted to define a 'perfect' food system:

A perfect food system delivers sufficient, affordable, sustainable and nutritious food for every human, reliably ensures adequacy for the future, respects others, both humans and nonhumans, causes minimal damage to the environment, acknowledges cultural diversity, and any resulting adverse consequences are fully mitigated with no unnecessary casualties.

We're a long way from this. Maybe we'll never get there. But we can edge closer. And we must. If we don't improve the way we eat, the dire consequences of the food system in its current guise will reverberate through all societies. The rise in rates of chronic diseases linked to lifestyle will not only place more demand on our already overly strained health services, but will be felt at an emotional level too. Suffering, by definition, is unpleasant. And it's not just those diagnosed with a disease who suffer: we all have sick loved ones and it's upsetting when we see them struggle – non-communicable diseases affect us all, even those of us who are healthy. This increase in emotional suffering will further exacerbate the mental health crisis, already made worse by poor dietary choices and the scourge of misinformation. In as little as a decade, the worsening effects of the environmental catastrophe will probably be felt in multiple regions across the world. Although many of you reading this will probably live in more affluent nations – providing some protection from the direct impacts of climate change – its indirect effects will affect us all. Fleeing from their homelands where life has become intolerable, migrants will make a beeline for places like Europe, Canada, Australia and the US, which means the effects of population density and the sharing of resources will become more challenging, increasing the chance of conflict. Populism can only protect borders for so long.

But we can change the food system, together. Every action can have an impact. And changing the way you eat, as we've seen, needn't be hard. Simply by making a few improvements, you'll be surprised how rapidly you can reap the benefits. To eat contemplatively is to consider what you're eating. The pillar system is straightforward: five fundamental considerations to keep in mind each time you shop, prepare a meal or reach for a snack. Become aware of the tantalising food cues and notice that your tastes, impulses, traditions, values and anxieties are telling you to chase foods that won't always be the wisest choices for your health, others or the environment. Contemplative nutrition involves embracing not fearing food innovations, but the right technologies: ones that enhance the nutritional quality of diets and minimise the impact of what you eat on others and the environment.

Be a little more 'we-focused' and slightly less 'me-focused'. Through contemplative eating we can impact on each other. Look to friends and family for support. There are huge benefits from socialising around food. And there are other ways to be part of a community: social clubs, gyms, dancing, coffee mornings and going for walks with friends can be great ways to exercise and meet people. By changing what you eat through contemplative nutrition, you can influence others. Maybe, even, you can prepare meals together.

By following the guidelines above, you can contribute to a food system that moves in the direction of 'perfect'. They are rational, reason-based, realistic, straightforward, enjoyable and, crucially, will make a difference. However, merely acknowledging that they make sense won't be enough: you have to commit to change what you eat even if it's just a little. But contemplation needn't be hard; after all, we don't want to have to think about our food choices *too* much.

We can look to mindfulness for help. Being introspective helps us address our food anxieties and become more aware of the tight grip that food cues have on us. By acknowledging that we're biological and cultural creatures, we notice that we don't have to succumb to the hedonic pulls and we can continue to take pleasure in eating while making better choices. I'm not necessarily talking about formal meditation practice, although 10 minutes a day would probably be beneficial, but I encourage you to pause and consider – *contemplate* – the food choices you make and their effects on your own health and wellbeing, the environment and others.

Contemplative nutrition is an evidence-based dietary approach for living in the Anthropocene. It's a dietary philosophy with rational decision-making at its core. It's a strategy that helps to address the many food-related hurdles before us, enabling you, your family, your friends and community, and the world in which we all live to flourish.

Contemplative nutrition is a mindset.

Appendix

Contemplating My Own Contemplative Diet

Contemplative nutrition evolved from my own dietary strategy. I limit the consumption of animal products to around 10 per cent of my calories and I try to eat only red meat, poultry and eggs that I can be reasonably sure have not come from intensive farming. My diet is high in protein, essential fats and fibre-rich carbs, and is based on a range of foods, including plenty of fruit, veg, pulses, nuts, seeds, grains and tubers.

One week's animal-derived calories are around 2,500 and might look something like this:

  1–1.5 pints of skimmed milk
  500 g yoghurt
  2–3 x 120 g fish (oily or white)
  1–2 x 100 g red meat, chicken or duck
  80 g cheese
  3–6 eggs

All the red meat and poultry we buy are from free-roaming farmed animals; my wife and I buy in bulk every four to six months. Eggs are bought from local farms where the chickens are free-roaming. I also eat out maybe as much as three or four times a week, either for work-related meetings or social engagements. Food choice is easy and meals are enjoyable. I'm not a fussy eater, and, although I have a limited flair for creative cooking, fortunately, my wife is an innovative cook and little goes to waste. Complete foods, such as Huel, are naturally a regular part of my diet, and this makes things considerably easier. Most days, I fast for around 16 hours and consume my food in the remaining eight-hour window. Although I don't follow a meal plan and don't routinely count calories or measure portion sizes, the following is a rough example of what I might have, though this varies considerably from week to week:

> **12 pm** – Pulses and/or oily fish and/or nuts + large salad + extra-virgin olive oil + rice/granary bread + natural peanut butter or Huel
>
> **2pm** – Huel
>
> **4pm** – Nuts + seeds + dried coconut + berries or Huel
>
> **7pm** – Pulse-based stew + veg + potatoes/rice or lean meat/fish + pulses + veg + potatoes/rice
>
> **8pm** – Nuts + seeds + dried coconut or wholegrain cereal + natural yoghurt or skimmed milk + 2 satsumas or berries

References

Chapter 1: An Imperfect Food System

1. Food and Agriculture Organization of the United Nations (2021) *The State of Food Security in the World 2021: The World Is at a Critical Juncture.* Available at: https://www.fao.org/state-of-food-security-nutrition (Accessed: 23 July 2024).
2. UN Environment Programme (2022) *Think Eat Save.* Available at: https://www.unep.org/resources/publication/food-waste-index-report-2024 (Accessed: 23 July 2024).
3. Harari, Y. N. (2011) *Sapiens: A Brief History of Humankind*, London: Penguin, p. 58.
4. Ibid.
5. Roser, M. (2013) 'Employment in Agriculture', *Our World in Data.* Available at: https://ourworldindata.org/employment-in-agriculture (Accessed: 24 July 2024).
6. Ehrlich, P. R. (1968) *The Population Bomb* (19th edn, 1970), New York: Ballantine Books, p. 96.
7. Examples include: (a) Levine, R. (2015) 'My Hero is Norman Borlaug, and Here's Why', *Entomology Today*, 12 October. Available at: https://entomologytoday.org/2015/10/12/my-hero-is-norman-borlaug-and-heres-why/ (Accessed: 23 July 2024); (b) Macaray, D. (2013) 'The Man Who Saved a Billion Lives', *HuffPost*, 15 October. Available at: https://www.huffpost.com/entry/the-man-who-saved-a-billi_b_4099523 (Accessed: 23 July 2024); (c) Easterbrook, G. (1997) 'Forgotten Benefactor of Humanity', *The Atlantic*, January. Available at: https://www.theatlantic.com/magazine/archive/1997/01/forgotten-benefactor-of-humanity/306101/ (Accessed: 23 July 2024); (d) Gates, B. (2021) *How to Avoid a Climate Disaster: The*

Solutions We Have and the Breakthroughs We Need, London: Allen Lane, pp. 114–15.

8. Ray, C. C. (2015) 'A Decline in the Nutritional Value of Crops', *The New York Times*, 12 September. Available at: https://www.nytimes.com/2015/09/15/science/a-decline-in-the-nutritional-value-of-crops.html (Accessed: 14 June 2024).

9. Mayer, A.-M. B. et al. (2021) 'Historical Changes in the Mineral Content of Fruit and Vegetables in the UK from 1940 to 2019: A Concern for Human Nutrition and Agriculture', *International Journal of Food Sciences and Nutrition*, 73(3), 315–26.

10. Davis, D. R. et al. (2004) 'Changes in USDA Food Composition Data for 43 Garden Crops, 1950 to 1999', *Journal of the American College of Nutrition*, 23(6), 669–82.

11. Varshney, V. (2017) 'Food Basket in Danger', DownToEarth, 28 February. Available at: https://www.downtoearth.org.in/news/health/food-basket-in-danger-57079 (Accessed: 23 July 2024).

12. Jarrell, W. M. and Beverly, R. B. (1981) 'The Dilution Effect in Plant Nutrition Studies', *Advances in Agronomy*, 34, 197–224.

13. Mariem, S. B. et al. (2020) 'Assessing the Evolution of Wheat Grain Traits During the Last 166 Years Using Archived Samples', *Scientific Reports*, 10(1), 21828.

14. Myers, S. et al. (2014) 'Increasing CO_2 Threatens Human Nutrition', *Nature*, 510, 139–42.

15. Feng, Z. et al. (2015) 'Constraints to Nitrogen Acquisition of Terrestrial Plants Under Elevated CO_2', *Global Change Biology*, 21(8), 3152–68.

16. Montgomery, D. R. et al. (2022) 'Soil Health and Nutrient Density: Preliminary Comparison of Regenerative and Conventional Farming', *PeerJ*, 10, e12848.

17. (a) Welch, R. M. and Graham, R. D. (2004) 'Breeding for Micronutrients in Staple Food Crops from a Human Nutrition Perspective', *Journal of Experimental Botany*, 55(396), 353–64; (b) High Level Panel of Experts on Food Security and Nutrition (2016) *Project Team for the Report on Nutrition and Food Systems*. Available at: https://www.fao.org/fileadmin/user_upload/hlpe/hlpe_documents/PT_Nutrition/Docs/HLPE_Nutrition_Project-Team_9_May_2016.pdf (Accessed: 23 July 2024).

18. Ritchie, H. and Roser, M. (2017) 'Micronutrient Deficiency', *Our World in Data* August. Available at: https://ourworldindata.org/micronutrient-deficiency (Accessed: 23 July 2024).

19. White, P. J. and Broadley, M. R. (2005) 'Historical Variation in the Mineral Composition of Edible Horticultural Products', *The Journal of Horticultural Science and Biotechnology*, 80(6), 660–7.

20. Eustachio Colombo, P. et al. (2021) 'Pathways to "5-a-day": Modeling the Health Impacts and Environmental Footprints of Meeting the Target for Fruit and Vegetable Intake in the United Kingdom', *The American Journal of Clinical Nutrition*, 114(2), 530–9.

Chapter 2: Our Drive to Eat

1. Harris, M. (1985) *The Sacred Cow and the Abominable Pig: Riddles of Food and Culture*, New York: Simon & Schuster, p. 236.

2. Burgoine, T. et al. (2014) 'Associations between Exposure to Takeaway Food Outlets, Takeaway Food Consumption, and Body Weight in Cambridgeshire, UK: Population Based, Cross Sectional Study', *BMJ*, 348.

3. Zion Market Research (2019) 'Global Industry Trends in Fast Food Market Size & Share Will Surpass USD 690.80 Billion by 2022', *GlobeNewswire*, 12 July. Available at: http://www.globenewswire.com/news-release/ 2019/07/12/1882007/0/en/Global-Industry-Trends-in-Fast-Food-Market-Size-Share-Will-Surpass-USD-690-80-Billion-by-2022.html (Accessed: 23 July 2024).

4. (a) Middaugh, A. L. et al. (2012) 'Few Associations Between Income and Fruit and Vegetable Consumption', *Journal of Nutrition Education and Behavior*, 44(3), 196–203; (b) Grimm, K. A. et al. (2012) 'Household Income Disparities in Fruit and Vegetable Consumption by State and Territory: Results of the 2009 Behavioral Risk Factor Surveillance System', *Journal of the Academy of Nutrition and Dietetics*, 112(12), 2014–21; (c) Miller, V. et al. (2016) 'Availability, Affordability, and Consumption of Fruits and Vegetables in 18 Countries Across Income Levels: Findings from the Prospective Urban Rural Epidemiology (PURE) Study', *The Lancet Global Health*, 4(10), e695–703.

5. Arias-Carrión, O. et al. (2010) 'Dopaminergic Reward System: A Short Integrative Review', *International Archives of Medicine*, 3, 24.

6. Chase, H. W. and Clark, L. (2010) 'Gambling Severity Predicts Midbrain Response to Near-Miss Outcomes', *Journal of Neuroscience*, 30(18), 6180–7.

7. Arias-Carrión et al., 'Dopaminergic Reward System', 24.

8. Hsu, T. M. et al. (2018) 'Parallels and Overlap: The Integration of Homeostatic Signals by Mesolimbic Dopamine Neurons', *Frontiers in Psychiatry*, 9, 410.

9. Zhang, Y. et al. (1995) 'Positional Cloning of the Mouse Obese Gene and Its Human Homologue', *Nature*, 372(6505), 425–32.

10. (a) Nieto-Vazquez, I. et al. (2008) 'Insulin Resistance Associated to Obesity: The Link TNF-alpha', *Archives of Physiology and Biochemistry*, 114(3), 183–94; (b) Ramírez, S. and Claret, M. (2015) 'Hypothalamic ER Stress: A Bridge between Leptin Resistance and Obesity', *FEBS Letters*, 14, 1678–87; (c) Jenkinson, A. (2020) *Why We Eat (Too Much): The New Science of Appetite*, London: Penguin, p. 95.

11. Kojima, M. et al. (1999) 'Ghrelin Is a Growth-Hormone-Releasing Acylated Peptide from Stomach', *Nature*, 402(6762), 656–60.

12. Dickson, S. L. et al. (1993) 'Systemic Administration of Growth Hormone-Releasing Peptide Activates Hypothalamic Arcuate Neurons', *Neuroscience*, 53(2), 303–6.

13. Nakazato, M. et al. (2001) 'A Role for Ghrelin in the Central Regulation of Feeding', *Nature*, 409, 194–8.

14. Cummings, D. E. et al. (2002) 'Plasma Ghrelin Levels After Diet-Induced Weight Loss or Gastric Bypass Surgery', *New England Journal of Medicine*, 346(21), 1623–30.

15. (a) Batterham, R. L. et al. (2007) 'PYY Modulation of Cortical and Hypothalamic Brain Areas Predicts Feeding Behaviour in Humans', *Nature*, 450(7166), 106–9; (b) Karra, E. et al. (2009) 'The Role of Peptide YY in Appetite Regulation and Obesity', *Journal of Physiology*, 587(1), 19–25.

16. Cawthorn, C. R. and de La Serre, C. B. (2021) 'The Critical Role of CCK in the Regulation of Food Intake and Diet-Induced Obesity', *Peptides*, 138, 170492.

17. Hayashi, D. et al. (2023) 'What Is Food Noise? A Conceptual Model of Food Cue Reactivity', *Nutrients*, 15(22), 4809.

18. Barrera, J. G. et al. (2011) 'GLP-1 and Energy Balance: An Integrated Model of Short-Term and Long-Term Control', *Nature Reviews: Endocrinology*, 7(9), 507–16.

19. (a) National Institute for Health and Care Excellence (2023) *Semaglutide for Managing Overweight and Obesity. Technology Appraisal Guidance. TA875. Recommendations.* Available at: https://www.nice.org.uk/guidance/ta875/chapter/1-Recommendations (Accessed: 25 July 2024); (b) US Food & Drug Administration (2024) *Medications Containing Semaglutide*

References

Marketed for Type 2 Diabetes or Weight Loss. Available at: https://www.fda. gov/drugs/postmarket-drug-safety-information-patients-and-providers/ fdas-concerns-unapproved-glp-1-drugs-used-weight-loss (Accessed: 16 October 2024).

20. Young, L. J. (2024) 'Ozempic Quiets Food Noise in the Brain – But How?', *Scientific American*, 25 June. Available at: https://www.scientificamerican. com/article/ozempic-quiets-food-noise-in-the-brain-but-how/ (Accessed: 7 August 2024).

21. Phoenix Veterinary Center (2020) 'How Powerful Is a Dog's Nose?', 23 July. Available at: https://phoenixvetcenter.com/blog/214731-how-powerful-is-a-dogs-nose (Accessed: 23 July 2024).

22. Hall, J. E. and Hall, M. E. (2021) *Guyton & Hall Textbook of Medical Physiology* (14th edn), Edinburgh: Elsevier, Chapter 54.

23. Ibid.

24. (a) Keast, R. S. and Costanzo, A. (2015) 'Is Fat the Sixth Taste Primary? Evidence and Implications', *Flavour*, 4, 5; (b) Cordelia, A. et al. (2015) 'Oleogustus: The Unique Taste of Fat', *Chemical Senses*, 40(7), 507–16.

25. (a) Prescott, J. and Stevenson, R. J. (1995) 'Pungency in Food Perception and Preference', *Food Reviews International*, 11(4), 665–98; (b) Ludy, M.-J. and Mattes, R. D. (2012) 'Comparison of Sensory, Physiological, Personality, and Cultural Attributes in Regular Spicy Food Users and Non-Users', *Appetite*, 58(1), 19–27; (c) Torrico, D. D. et al. (2019) 'Cross-Cultural Effects of Food Product Familiarity on Sensory Acceptability and Non-Invasive Physiological Responses of Consumers', *Food Research International*, 115, 439–50.

26. Wadhera, D. and Capaldi-Phillips, E. D. (2014) 'A Review of Visual Cues Associated with Food on Food Acceptance and Consumption', *Eating Behaviors*, 15(1), 132–43.

Chapter 3: Contemplating Nutrition Misinformation

1. Statista (2024) *Number of Social Media Users Worldwide from 2017 to 2027 (in Billions)*. Available at: https://www.statista.com/statistics/278414/ number-of-worldwide-social-network-users/ (Accessed: 24 July 2024).

2. Oracle (2022) '37% of Consumers Trust Social Media Influencers over Brands', *PR Newswire*, 3 May. Available at: https://www.prnewswire.com/ news-releases/37-of-consumers-trust-social-media-influencers-over-brands-301538111.html (Accessed: 24 July 2024).

3. (a) Schouten, A. P. et al. (2020) 'Celebrity vs. Influencer Endorsements in Advertising: The Role of Identification, Credibility, and Product-Endorser Fit', *International Journal of Advertising*, 39(2), 258–81; (b) Chung, A. et al. (2021) 'Adolescent Peer Influence on Eating Behaviors via Social Media: Scoping Review', *Journal of Medical Internet Research*, 23(6), e19697.

4. Alwafi, H. et al. (2022) 'The Impact of Social Media Influencers on Food Consumption in Saudi Arabia, a Cross-Sectional Web-Based Survey', *Journal of Multidisciplinary Healthcare*, 15, 2129–39.

5. Coates, A. E. et al. (2019) 'Social Media Influencer Marketing and Children's Food Intake: A Randomized Trial', *Pediatrics*, 143(4), e20182554.

6. Spence, C. et al. (2016) 'Eating with Our Eyes: From Visual Hunger to Digital Satiation', *Brain and Cognition*, 110, 53–63.

7. Forrest, A. (2019) 'Social Media Influencers Give Bad Diet and Fitness Advice Eight Times Out of Nine, Research Reveals', *Independent*, 30 April. Available at: https://www.independent.co.uk/news/health/social-media-weight-loss-diet-twitter-influencers-bloggers-glasgow-university-a8891971.html (Accessed: 24 July 2024).

8. Denniss, E. et al. (2024) '#Fail: The Quality and Accuracy of Nutrition-Related Information by Influential Australian Instagram Accounts', *International Journal of Behavioral Nutrition and Physical Activity*, 21(1), 16.

9. Turel, O. et al. (2014) 'Examination of Neural Systems Sub-Serving Facebook "Addiction"', *Psychological Reports*, 115(3), 675–95.

10. Taylor, C. (2023) 'Got to Have It: The Dangers of Social Media Impulse Buying', *Reuters*, 28 September. Available at: https://www.reuters.com/technology/got-have-it-dangers-social-media-impulse-buying-2023-09-28/ (Accessed: 27 July 2024).

11. Flaminjoy (2023) 'Influencer Marketing in the Food Industry: The Who, the What, and the ROI', 4 January. Available at: https://www.flaminjoy.com/blog/influencer-marketing-food-industry-roi/ (Accessed: 24 July 2024).

12. Ibid.

13. Reed, K. E. et al. (2021) 'Neither Soy nor Isoflavone Intake Affects Male Reproductive Hormones: An Expanded and Updated Meta-Analysis of Clinical Studies', *Reproductive Toxicology*, 100, 60–7.

14. Jingnan, H. (2023) 'How a Conspiracy Theory about Eating Bugs Made Its Way to International Politics', NPR, 31 March. Available at: https://www.npr.org/2023/03/31/1167550482/how-a-conspiracy-theory-about-eating-bugs-made-its-way-to-international-politics (Accessed: 25 July 2024).

References

15. Vosoughi, S. et al. (2018) 'The Spread of True and False News Online', *Science*, 359, 1146–51.
16. (a) Senz, K. (2021) 'Outrage Spreads Faster on Twitter: Evidence from 44 News Outlets', *Harvard Business School Working Knowledge*, 13 July. Available at: https://hbswk.hbs.edu/item/hate-spreads-faster-on-twitter-evidence-from-44-news-outlets (Accessed: 25 July 2024); (b) Robertson, C. E. et al. (2023) 'Negativity Drives Online News Consumption', *Nature Human Behaviour*, 7, 812–22.
17. Galmiche, M. et al. (2019) 'Prevalence of Eating Disorders over the 2000–2018 Period: A Systematic Literature Review', *The American Journal of Clinical Nutrition*, 109(5), 1402–13.
18. Jenkins, Z. M. et al. (2020) 'A Comparison of Eating Disorder Symptomatology, Psychological Distress and Psychosocial Function Between Early, Typical and Later Onset Anorexia Nervosa', *Journal of Eating Disorders*, 8, 56.
19. BBC (2018) 'The Shocking Amount of Sugar Hiding in Your Food – BBC', YouTube. Available at: https://youtu.be/eKQWFJmCWZE (Accessed: 24 July 2024).
20. Nichols, T. (2017) *The Death of Expertise*, New York: Oxford University Press, p. 245.
21. Richie, S. (2020) *Science Fictions*, London: The Bodley Head, p. 169.
22. Association for Nutrition (2024) *Association for Nutrition*. Available at: https://www.associationfornutrition.org/ (Accessed: 25 July 2024).
23. British Dietetic Association (2024) *What Is a Dietitian?* Available at: https://www.bda.uk.com/about-dietetics/what-is-dietitian.html (Accessed: 25 July 2024).
24. Academy of Nutrition and Dietetics (2024) *About RDNs and NDTRs.* Eatright. Available at: https://www.eatright.org/about-rdns-and-ndtrs (Accessed: 25 July 2024).
25. 'Regulation (EC) No. 1924/2006 of the European Parliament and of the Council of 20 December 2006 on Nutrition and Health Claims Made on Foods' (2006) *Official Journal L404*, 9–25.
26. Ruxton, C. and Ashwell, M. (2023) 'Dietitians' and Nutritionists' Knowledge and Views on Aspects of Health Claims Regulation in the UK: Do We Inadvertently Shoot the Messenger?', *Nutrition Bulletin*, 48, 216–26.
27. Harris, J. L. et al. (2021) *Fast Food Facts 2021. Fast Food Advertising: Billions in Spending, Continued High Exposure by Youth*, UConn Rudd Center for Food Policy & Obesity. Available at: https://media.ruddcenter.

uconn.edu/wp-content/uploads/sites/2909/2024/06/FACTS2021.pdf (Accessed: 29 July 2024).

28. Roe, B. et al. (1999) 'The Impact of Health Claims on Consumer Search and Product Evaluation Outcomes: Results from FDA Experimental Data', *Journal of Public Policy & Marketing*, 18(1), 89–105.

29. Ibid.

30. Ibid.

Chapter 4: Contemplating Food and Our Physical Health

1. Alpha-Tocopherol, Beta Carotene Cancer Prevention Study Group (1994) 'The Effect of Vitamin E and Beta Carotene on the Incidence of Lung Cancer and Other Cancers in Male Smokers', *New England Journal of Medicine*, 330(15), 1029–35.

2. Our World in Data (n.d.) *Share of Consumer Expenditure Spent on Food vs. Total Consumer Expenditure, 2021.* Available at: https://ourworldindata. org/grapher/food-expenditure-share-gdp?time=2021 (Accessed: 24 July 2024).

3. GBD 2017 Diet Collaborators (2019) 'Health Effects of Dietary Risks in 195 Countries, 1990–2017: A Systematic Analysis for the Global Burden of Disease Study 2017', *The Lancet*, 393(10184), 1958–72.

4. Ibid.

5. GBD 2015 Obesity Collaborators (2017) 'Health Effects of Overweight and Obesity in 195 Countries over 25 Years', *New England Journal of Medicine*, 377, 13–27.

6. Centers for Disease Control and Prevention (2024) *Adult Obesity Facts.* Available at: https://www.cdc.gov/obesity/php/data-research/adult-obesity-facts.html?CDC_AAref_Val=https://www.cdc.gov/obesity/data/adult.html (Accessed: 24 July 2024).

7. Ward, Z. J. et al. (2021) 'Association of Body Mass Index with Health Care Expenditures in the United States by Age and Sex', *PLOS ONE*, 16(3), e0247307

8. Food and Agriculture Organization of the United Nations et al. (2019) *The State of Food Security and Nutrition in the World 2019: Safeguarding Against Economic Slowdowns and Downturns.* Available at: https://docs. wfp.org/api/documents/WFP-0000106760/download/?_ga= 2.13038135.1476546118.1679995260-755977012.1679995260 (Accessed: 24 July 2024)

References

9. Health Poverty Action (n.d.) *Food & Nutrition*. Available at: https://www.
   healthpovertyaction.org/how-poverty-is-created/essentials-for-health/
   food-and-nutrition/?gclid=Cj0KCQjw0emHBhC1ARIsAL1QGNezLddhN
   Fv-c_uWRCm86Za_Mc51DD941HwTMI-
   9dMf4lQvcO5XmdKUaAtnUEALw_wcB (Accessed: 24 July 2024).

10. James, W. P. (2008) 'WHO Recognition of the Global Obesity Epidemic',
    International Journal of Obesity, 32(S7), s120–6.

11. Eknoyan, G. (2008) 'Adolphe Quetelet (1796–1874) – The Average Man
    and Indices of Obesity', *Nephrology Dialysis Transplantation*, 23(1), 47–51.

12. Blackburn, H. and Jacobs, D. (2014) 'Commentary: Origins and Evolution
    of Body Mass Index (BMI): Continuing Saga', *International Journal of
    Epidemiology*, 43(3), 665–9.

13. Mishra, K. and Floegel-Shetty, A. (2023) 'What's Wrong with Overreliance
    on BMI?', *AMA Journal of Ethics*, 25(7), e469–71.

14. NHS (2023) *Treatment: Obesity*. Available at: https://www.nhs.uk/
    conditions/obesity/treatment/ (Accessed: 25 July 2024).

15. Rosen, H. (2014) 'Is Obesity a Disease or a Behavior Abnormality? Did the
    AMA Get It Right?' *Missouri Medicine*, 111(2), 1048.

16. (a) Stefan, N. et al. (2013) 'Metabolically Healthy Obesity: Epidemiology,
    Mechanisms, and Clinical Implications', *The Lancet Diabetes &
    Endocrinology*, 1(2), 152–62; (b) April-Sanders, A. K. and Rodriguez, C. J.
    (2021) 'Metabolically Healthy Obesity Redefined', *JAMA Network Open*,
    4(5), e218860.

17. Steele, M. and Finucane, F. M. (2023) 'Philosophically, Is Obesity Really a
    Disease?', *Obesity Reviews*, 24(8), e13590.

18. (a) Min, J. et al. (2013) 'Variation in the Heritability of Body Mass Index
    Based on Diverse Twin Studies: A Systematic Review', *Obesity Reviews*,
    14(11), 871–82; (b) Silventoinen, K. et al. (2016) 'Genetic and
    Environmental Effects on Body Mass Index from Infancy to the Onset of
    Adulthood: An Individual-Based Pooled Analysis of 45 Twin Cohorts
    Participating in the COllaborative Project of Development of
    Anthropometrical Measures in Twins (CODATwins) Study', *American
    Journal of Clinical Nutrition*, 104(2), 371–9; (c) Plomin, R. (2018) *Blueprint:
    How DNA Makes Us Who We Are*, London: Penguin, pp. 27–9.

19. Jenkinson, A. (2020) *Why We Eat (Too Much): The New Science of Appetite*,
    London: Penguin, pp. 13–16.

20. Miller, P. E. et al. (2015) 'The United States Food Supply Is Not Consistent
    with Dietary Guidance: Evidence from an Evaluation Using the Healthy

Eating Index-2010', *Journal of the Academy of Nutrition and Dietetics*, 115(1), 95–100.

21. Hetherington, A. W. and Ranson, S. W. (1940) 'Hypothalamic Lesions and Adiposity in the Rat', *The Anatomical Record*, 78(2), 149–72.

22. Rothwell, N. J. and Stock, M. J. (1979) 'Regulation of Energy Balance in Two Models of Reversible Obesity in the Rat', *Journal of Comparative and Physiological Psychology*, 93(6), 1024–34.

23. Keys, A. et al. (1950) *The Biology of Human Starvation*, Minneapolis, MN: University of Minnesota Press.

24. Speakman, J. R. et al. (2011) 'Set Points, Settling Points and Some Alternative Models: Theoretical Options to Understand How Genes and Environments Combine to Regulate Body Adiposity', *Disease Models & Mechanisms*, 4(6), 733–45.

25. (a) Jenkinson, *Why We Eat (Too Much)*, pp. 28–30; (b) Keesey, R. E. and Hirvonen, M. D. (1997) 'Body Weight Set-Points: Determination and Adjustment', *Journal of Nutrition*, 127(9), 1875–83s.

26. Jenkinson, *Why We Eat (Too Much)*, p. 28.

27. Ng, S. W. and Popkin, B. M. (2012) 'Time Use and Physical Activity: A Shift Away from Movement across the Globe', *Obesity Reviews*, 13(8), 659–80.

28. Nakamura, S. et al. (2016) 'Gene–Environment Interactions in Obesity: Implication for Future Applications in Preventive Medicine', *Journal of Human Genetics*, 61, 317–22.

29. Müller, M. J. et al. (2018) 'Recent Advances in Understanding Body Weight Homeostasis in Humans', *F1000 Research*, 7, F1000 Faculty Rev-1025.

30. Pearce, M. et al. (2018) 'Weight Gain in Mid-Childhood and Its Relationship with the Fast Food Environment', *Journal of Public Health*, 40(2), 237–44.

31. Keijer, J. et al. (2014) 'Nutrigenomics of Body Weight Regulation: A Rationale for Careful Dissection of Individual Contributors', *Nutrients*, 6, 4531–51.

32. Wolrich, J. (2021) *Food Isn't Medicine: Challenge Nutrib*llocks & Escape the Diet Trap*, London: Penguin Random House, pp. 51–6, pp. 244–54.

33. (a) Flegal, K. M. et al. (2013) 'Association of All-Cause Mortality with Overweight and Obesity Using Standard Body Mass Index Categories: A Systematic Review and Meta-Analysis', *JAMA*, 309(1), 71–82; (b) Bhaskaran, K. et al. (2018) 'Association of BMI with Overall and Cause-Specific Mortality: A Population-Based Cohort Study of 3.6 Million Adults in the UK', *The Lancet: Diabetes & Endocrinology*, 6(12), 944–53.

34. (a) Kwon, Y. et al. (2017) 'Body Mass Index-Related Mortality in Patients with Type 2 Diabetes and Heterogeneity in Obesity Paradox Studies: A Dose-Response Meta-Analysis', *PLoS One*, 12(1), e0168247; (b) Carbone, S. et al. (2019) 'Obesity Paradox in Cardiovascular Disease: Where Do We Stand?', *Vascular Health and Risk Management*, 15, 89–100.

35. Aschwanden, C. (2024) 'People Who Are Fat and Healthy May Hold Keys to Understanding Obesity', *Scientific American*, 25 June. Available at: https://www.scientificamerican.com/article/people-who-are-fat-and-healthy-may-hold-keys-to-understanding-obesity/ (Accessed: 7 August 2024).

36. Examples: (a) Bogers, R. P. et al. (2007) 'Association of Overweight with Increased Risk of Coronary Heart Disease Partly Independent of Blood Pressure and Cholesterol Levels: A Meta-Analysis of 21 Cohort Studies Including More Than 300,000 Persons', *Archives of Internal Medicine*, 167(16), 1720–8; (b) Akil, L. and Ahmad, H. A. (2011) 'Relationships Between Obesity and Cardiovascular Diseases in Four Southern States and Colorado', *Journal of Health Care for the Poor and Underserved*, 22(S4), 61–72; (c) Barroso, T. A. et al. (2017) 'Association of Central Obesity with the Incidence of Cardiovascular Diseases and Risk Factors', *International Journal of Cardiovascular Science*, 30(5), 416–24; (d) Dwivedi, A. K. et al. (2020) 'Association Between Obesity and Cardiovascular Outcomes: Updated Evidence from Meta-Analysis Studies', *Current Cardiology Reports*, 22(4), 25.

37. Examples: (a) Da Silva, M. et al. (2018) 'Excess Body Weight, Weight Gain and Obesity-Related Cancer Risk in Women in Norway: The Norwegian Women and Cancer Study', *British Journal of Cancer*, 119, 646–56; (b) Bhaskaran, K. et al. (2014) 'Body-Mass Index and Risk of 22 Specific Cancers: A Population-Based Cohort Study of 5.24 Million UK Adults', *The Lancet*, 384(9945), 755–65.

38. Examples: (a) Bays, H. E. et al. (2007) 'The Relationship of Body Mass Index to Diabetes Mellitus, Hypertension and Dyslipidaemia: Comparison of Data from Two National Surveys', *International Journal of Clinical Practice*, 61(5), 737–47; (b) Ganz, M. L. et al. (2014) 'The Association of Body Mass Index with the Risk of Type 2 Diabetes: A Case-Control Study Nested in an Electronic Health Records System in the United States', *Diabetology & Metabolic Syndrome*, 6, 50; (c) Bell, J. A. et al. (2014) 'Metabolically Healthy Obesity and Risk of Incident Type 2 Diabetes: A Meta-Analysis of Prospective Cohort Studies', *Obesity Reviews*, 15(6), 504–15.

39. Examples: (a) Abdelaal, M. et al. (2017) 'Morbidity and Mortality Associated with Obesity', *Annals of Translational Medicine*, 5(7), 161; (b) Bhaskaran, K. et al. (2018) 'Association of BMI with Overall and Cause-Specific Mortality: A Population-Based Cohort Study of 3.6 Million Adults in the UK', *The Lancet: Diabetes & Endocrinology*, 6(12), 944–53.

40. Examples: (a) Vincent, H. K. et al. (2010) 'Obesity and Mobility Disability in the Older Adult', *Obesity Reviews*, 11(8), 568–79; (b) Houston, D. K. et al. (2009) 'Overweight and Obesity over the Adult Life Course and Incident Mobility Limitation in Older Adults: The Health, Aging and Body Composition Study', *American Journal of Epidemiology*, 169(8), 927–36.

41. Srivastava, S. et al. (2021) 'Interaction of Physical Activity on the Association of Obesity-Related Measures with Multimorbidity among Older Adults: A Population-Based Cross-Sectional Study in India', *BMJ Open*, 11, e050245.

42. Wolrich, *Food Isn't Medicine*, pp. 42–8.

43. Ibid., pp. 15–25.

44. O'Keefe Jr., J. H. and Cordain L. (2004) 'Cardiovascular Disease Resulting from a Diet and Lifestyle at Odds with Our Paleolithic Genome: How to Become a 21st-Century Hunter-Gatherer', *Mayo Clinic Proceedings*, 79(1), 101–8.

45. (a) Butovskaya, M. et al. (2017) 'Waist-to-Hip Ratio, Body-Mass Index, Age and Number of Children in Seven Traditional Societies', *Scientific Reports*, 7(1), 1622; (b) Pontzer, H. et al. (2018) 'Hunter-Gatherers as Models in Public Health', *Obesity Reviews*, 19(S1), 24–35.

46. Lopez, K. N. and Knudson, J. D. (2012) 'Obesity: From the Agricultural Revolution to the Contemporary Pediatric Epidemic', *Congenital Heart Disease*, 7(2), 189–99.

47. World Obesity (n.d.) *Prevalence of Obesity*. Available at: https://www.worldobesity.org/about/about-obesity/prevalence-of-obesity (Accessed: 24 July 2024)

48. Yeh, T. L. et al. (2019) 'The Relationship Between Metabolically Healthy Obesity and the Risk of Cardiovascular Disease: A Systematic Review and Meta-Analysis', *Journal of Clinical Medicine*, 8(8), 1228.

49. Pluckrose, H. and Lindsay, J. (2020) *Cynical Theories: How Activist Scholarship Made Everything About Race, Gender, and Identity – and Why This Harms Everybody*, Durham, NC: Pitchstone Publishing.

50. Wolrich, *Food Isn't Medicine*, pp. 44–8.

References

51. (a) Mensink, R. P. et al. (2003) 'Effects of Dietary Fatty Acids and Carbohydrates on the Ratio of Serum Total to HDL Cholesterol and on Serum Lipids and Apolipoproteins: A Meta-Analysis of 60 Controlled Trials', *American Journal of Clinical Nutrition*, 77(5), 1146–55; (b) Mozaffarian, D. et al. (2009) 'Health Effects of Trans-Fatty Acids: Experimental and Observational Evidence', *European Journal of Clinical Nutrition*, 63(S2), s5–21.

52. Mensink, R. P. and Katan, M. B. (1990) 'Effect of Dietary Trans Fatty Acids on High-Density and Low-Density Lipoprotein Cholesterol Levels in Healthy Subjects', *New England Journal of Medicine*, 323(7), 439–45.

53. Willett, W. C. (1993) 'Intake of Trans Fatty Acids and Risk of Coronary Heart Disease Among Women', *The Lancet*, 341(8845), 581–5.

54. Shapiro, S. (1997) 'Do Trans Fatty Acids Increase the Risk of Coronary Artery Disease? A Critique of the Epidemiologic Evidence', *American Journal of Clinical Nutrition*, 66(S4), 1011–17s.

55. Davidson, M. H. (2013) 'Omega-3 Fatty Acids: New Insights into the Pharmacology and Biology of Docosahexaenoic Acid, Docosapentaenoic Acid, and Eicosapentaenoic Acid', *Current Opinion in Lipidology*, 24(6), 467–74.

56. Yannios, T. (1999) *The Heart Disease Breakthrough*, New York: Wiley.

57. Horrocks, L. A. and Yeo, Y. K. (1999) 'Health Benefits of Docosahexaenoic Acid (DHA)', *Pharmacological Research*, 40(3), 211–25.

58. Linus Pauling Institute (2019) *Essential Fatty Acids*. Available at: https://lpi. oregonstate.edu/mic/other-nutrients/essential-fatty-acids (Accessed: 24 July 2024).

59. (a) Simopoulos, A. P. (2002) 'The Importance of the Ratio of Omega-6/ Omega-3 Essential Fatty Acids', *Biomedicine & Pharmacotherapy*, 56(8), 365–79; (b) Simopoulos, A. P. (2006) 'Evolutionary Aspects of Diet, the Omega-6/Omega-3 Ratio and Genetic Variation: Nutritional Implications for Chronic Diseases', *Biomedicine & Pharmacotherapy*, 60(9), 502–7.

60. Simopoulos, 'The Importance of the Ratio of Omega-6/Omega-3 Essential Fatty Acids'.

61. (a) Calder, P. C. (2010) 'Omega-3 Fatty Acids and Inflammatory Processes', *Nutrients*, 2(3), 355–74; (b) Wall, R. et al. (2010) 'Fatty Acids from Fish: The Anti-Inflammatory Potential of Long-Chain Omega-3 Fatty Acids', *Nutrition Reviews*, 68(5), 280–9.

62. (a) Simopoulos, A. P. (2008) 'The Omega-6/Omega-3 Fatty Acid Ratio, Genetic Variation, and Cardiovascular Disease', *Asia Pacific Journal of*

Clinical Nutrition, 17(S1), 131–4; (b) Julia, C. et al. (2013) 'Dietary Patterns and Risk of Elevated C-Reactive Protein Concentrations 12 Years Later', *British Journal of Nutrition*, 110(4), 747–54.

63. Simopoulos, A. P. (2006) 'Omega-6/Omega-3 Essential Fatty Acid Ratio and Chronic Disease', *Food Reviews International*, 20(1), 77–90.

64. Rett, B. S. and Whelan, J. (2011) 'Increasing Dietary Linoleic Acid Does Not Increase Tissue Arachidonic Acid Content in Adults Consuming Western-Type Diets: A Systematic Review', *Nutrition & Metabolism*, 8, 36.

65. Deckelbaum, R. J. et al. (2010) 'Dietary n-3 and n-6 Fatty Acids: Are There "Bad" Polyunsaturated Fatty Acids?', *Current Opinion in Clinical Nutrition and Metabolic Care*, 13(2), 123–4.

66. Values from (a) Nutritics (2024) Nutritics. Available at: https://www.nutritics.com/ (Accessed: 24 July 2024); (b) ExRx.net (2024) *Omega-3 Fatty Acid Content in Fish*. Available at: https://exrx.net/Nutrition/FishOmega3 (Accessed: 24 July 2024).

67. Nutrition Coalition (2020) 'Leading Scientists Agree: Current Limits on Saturated Fats No Longer Justified', 25 February. Available at: https://www.nutritioncoalition.us/news/saturated-fat-limit-not-justified (Accessed: 24 July 2024).

68. Yudkin, J. (1972) *Pure, White and Deadly: How Sugar Is Killing Us and What We Can Do to Stop It*, London: Davis-Poynter; reissue London: Penguin, 2012.

69. Kearns, C. E. et al. (2016) 'Sugar Industry and Coronary Heart Disease Research: A Historical Analysis of Internal Industry Documents', *JAMA Internal Medicine*, 176(11), 1680–5.

70. Stallones, R. A. (1980) 'Patterns of Health Risk: *Seven Countries. A Multivariate Analysis of Death and Coronary Heart Disease. Ancel Keys with 15 others. Harvard University Press, Cambridge, Mass., 1980. xvi, 382 pp., illus. $25. A Commonwealth Fund Book*', *Science*, 208(4448), 1138–9.

71. Examples: (a) Ginsberg, H. N. et al. (1994) 'Effects of Increasing Dietary Polyunsaturated Fatty Acids Within the Guidelines of the AHA Step 1 Diet on Plasma Lipid and Lipoprotein Levels in Normal Males', *Arteriosclerosis and Thrombosis: A Journal of Vascular Biology*, 14(6), 892–901; (b) Fuehrlein, B. S. et al. (2004) 'Differential Metabolic Effects of Saturated Versus Polyunsaturated Fats in Ketogenic Diets', *The Journal of Clinical Endocrinology and Metabolism*, 89(4), 1641–5; (c) Risérus, U. et al. (2009) 'Dietary Fats and Prevention of Type 2 Diabetes', *Progress in Lipid Research*, 48(1), 44–51; (d) Mozaffarian, D. et al. (2010) 'Effects on

References

Coronary Heart Disease of Increasing Polyunsaturated Fat in Place of Saturated Fat: A Systematic Review and Meta-Analysis of Randomized Controlled Trials', *PLoS Med*, 7(3), e1000252: (e) Schwab, U. et al. (2021) 'Dietary Fat Intakes and Cardiovascular Disease Risk in Adults with Type 2 Diabetes: A Systematic Review and Meta-Analysis', *European Journal of Nutrition*, 60(6), 3355–63.

72. Hamley, S. (2017) 'The Effect of Replacing Saturated Fat with Mostly n-6 Polyunsaturated Fat on Coronary Heart Disease: A Meta-Analysis of Randomised Controlled Trials', *Nutrition Journal*, 16, 30.

73. Richie, S. (2020) *Science Fictions*, London: The Bodley Head, p. 313.

74. Hamley, 'The Effect of Replacing Saturated Fat with Mostly n-6 Polyunsaturated Fat on Coronary Heart Disease'.

75. Cordain, L. et al. (2002) 'The Paradoxical Nature of Hunter-Gatherer Diets: Meat-Based, yet Non-Atherogenic', *European Journal of Clinical Nutrition*, 56(S1), s42–52.

76. (a) Guasch-Ferré, M. et al. (2019) 'Meta-Analysis of Randomized Controlled Trials of Red Meat Consumption in Comparison with Various Comparison Diets on Cardiovascular Risk Factors', *Circulation*, 139(15), 1828–45; (b) Wang, D. D. et al. (2023) 'Red Meat Intake and the Risk of Cardiovascular Diseases: A Prospective Cohort Study in the Million Veteran Program', *The Journal of Nutrition*, 154(3), 886–95.

77. (a) Scientific Advisory Committee on Nutrition (2019) *Saturated Fats and Health*. Available at: https://assets.publishing.service.gov.uk/government/uploads/system/uploads/attachment_data/file/814995/SACN_report_on_saturated_fat_and_health.pdf (Accessed: 24 July 2024); (b) Office of Disease Prevention and Health Promotion (2016) *Cut Down on Saturated Fats*. Available at: https://health.gov/sites/default/files/2019-10/DGA_Cut-Down-On-Saturated-Fats.pdf (Accessed: 24 July 2024).

78. Sowell, T. (1995) *The Vision of the Anointed: Self-Congratulation as a Basis for Social Policy*, New York: Basic Books.

79. (a) British Nutrition Foundation (2023) *Fibre*. Available at: https://www.nutrition.org.uk/nutritional-information/fibre/ (Accessed: 25 July 2024); (b) Kehoe, L. et al. (2023) 'Food and Nutrient Intakes and Compliance with Recommendations in School-Aged Children in Ireland: Findings from the National Children's Food Survey II (2017–2018) and Changes since 2003–2004', *British Journal of Nutrition*, 129(11), 2011–24; (c) Boyle, N. B. et al. (2024) 'Increasing Fibre Intake in the UK: Lessons from the

Danish Whole Grain Partnership', *British Journal of Nutrition*, 131(4), 672–85.

80. ASN Staff (2021) 'Most Americans Are Not Getting Enough Fiber in Our Diets', *American Society for Nutrition*, 9 June. Available at: https://nutrition.org/most-americans-are-not-getting-enough-fiber-in-our-diets/ (Accessed: 25 July 2024).

81. Examples: (a) Aune, D. et al. (2016) 'Whole Grain Consumption and Risk of Cardiovascular Disease, Cancer, and All Cause and Cause Specific Mortality: Systematic Review and Dose-Response Meta-Analysis of Prospective Studies', *BMJ (Clinical Research Ed.)*, 353, i2716; (b) Benisi-Kohansal, S. et al. (2016) 'Whole-Grain Intake and Mortality from All Causes, Cardiovascular Disease, and Cancer: A Systematic Review and Dose-Response Meta-Analysis of Prospective Cohort Studies', *Advances in Nutrition*, 7(6), 1052–65.

82. Examples: (a) Martínez-González, M. A. (2015) 'Benefits of the Mediterranean Diet: Insights from the PREDIMED Study', *Progress in Cardiovascular Diseases*, 58(1), 50–60; (b) Guasch-Ferré, M. and Willett, W. C. (2021) 'The Mediterranean Diet and Health: A Comprehensive Overview', *Journal of Internal Medicine*, 290, 549–66; (c) Finicelli, M. et al. (2022) 'The Mediterranean Diet: An Update of the Clinical Trials', *Nutrients*, 14(14), 2956.

83. Sweney, M. (2021) 'Coca-Cola's Ronaldo Fiasco Highlights Risk to Brands in Social Media Age', *The Guardian*, 18 June. Available at: https://www.theguardian.com/media/2021/jun/18/coca-colas-ronaldo-fiasco-highlights-risk-to-brands-in-social-media-age (Accessed: 24 July 2024)

84. (a) Cooper, R. et al. (2012) 'Creatine Supplementation with Specific View to Exercise/Sports Performance: An Update', *Journal of the International Society of Sports and Nutrition*, 9(1), 33; (b) Mielgo-Ayuso, J. et al. (2019) 'Effects of Creatine Supplementation on Athletic Performance in Soccer Players: A Systematic Review and Meta-Analysis', *Nutrients*, 11(4), 757.

85. (a) RheoBlair.com (n.d.) *Rheo Blair*. Available at: https://www.rheoblair.com/ (Accessed: 24 July 2024); (b) Schinetsky, R. (2018) 'The Complete History of Protein Powder', *Tiger Fitness*, 2 October. Available at: https://www.tigerfitness.com/blogs/supplements/the-complete-history-of-protein-powder (Accessed: 24 July 2024).

86. (a) Salque, M. et al. (2013) 'Earliest Evidence for Cheese Making in the Sixth Millennium BC in Northern Europe', *Nature*, 493, 522–5; (b) McClure, S. B. et al. (2018) 'Fatty Acid Specific δ13C Values Reveal Earliest

References

Mediterranean Cheese Production 7,200 years ago', *PLoS One*, 13(9), e0202807.

87. (a) Mackenzie, C. (2018) 'This History of Protein Supplements', *Healthy for Men*, 3 July. Available at: https://www.healthyformen.com/history-protein-supplements/ (Accessed: 29 July 2024); (b) Orangefit (2021) 'The History of the Protein Shake', 24 September. Available at: https://www.orangefit.eu/fitblog/nutrition/the-history-of-the-protein-shake (Accessed: 29 July 2024).

88. Fortune Business Insights (2021) *Plant Based Protein Supplements Market Size, Share, & COVID-19 Impact Analysis*. Available at: https://www.fortunebusinessinsights.com/industry-reports/plant-based-protein-supplements-market-100082 (Accessed: 24 July 2024).

89. Grand View Research (2022) 'Protein Supplements Market Size Worth $10.80 Billion by 2030', November. Available at: https://www.grandviewresearch.com/press-release/global-protein-supplements-market (Accessed: 24 July 2024).

90. Kale, S. (2021) 'Muscles and Methane: How Protein Became the Food Industry's Biggest Craze', *The Guardian*, 15 September. Available at: https://www.theguardian.com/food/2021/sep/15/muscles-and-methane-how-protein-became-the-food-industrys-biggest-craze (Accessed: 24 July 2024).

91. Heffernan, C. (2016) 'Soy, Science and Selling: Bob Hoffman's Hi-Proteen Powder', *Physical Culture Study*, 15 June. Available at: https://physicalculturestudy.com/2016/06/15/soy-science-and-selling-bob-hoffmans-hi-proteen-powder/ (Accessed: 24 July 2024).

92. Morton, R. W. et al. (2018) 'A Systematic Review, Meta-Analysis and Meta-Regression of the Effect of Protein Supplementation on Resistance Training-Induced Gains in Muscle Mass and Strength in Healthy Adults', *British Journal of Sports Medicine*, 52, 376–84.

93. British Nutrition Foundation (2020) *Sport and Exercise: Eating Well for Exercise*. https://www.nutrition.org.uk/putting-it-into-practice/keeping-active/nutrition-for-sports-and-exercise/ (Accessed: 24 July 2024).

94. Dhillon, J. et al. (2016) 'The Effects of Increased Protein Intake on Fullness: A Meta-Analysis and Its Limitations', *Journal of the Academy of Nutrition and Dietetics*, 116(6), 968–83.

95. (a) Anthony, J. C. et al. (2000) 'Orally Administered Leucine Stimulates Protein Synthesis in Skeletal Muscle of Postabsorptive Rats in Association with Increased eIF4F Formation', *Journal of Nutrition*, 130(2), 139–45; (b)

Anthony, J. C. et al. (2000) 'Leucine Stimulates Translation Initiation in Skeletal Muscle of Postabsorptive Rats via a Rapamycin-Sensitive Pathway', *Journal of Nutrition*, 130(10), 2413–19; (c) Blomstrand, E. et al. (2006) 'Branched-Chain Amino Acids Activate Key Enzymes in Protein Synthesis After Physical Exercise', *Journal of Nutrition*, 136(S1), 269–73s.

96. Wilson, G. J. et al. (2008) 'Effects of Beta-Hydroxy-Beta-Methylbutyrate (HMB) on Exercise Performance and Body Composition Across Varying Levels of Age, Sex, and Training Experience: A Review', *Nutrition & Metabolism*, 5, 1.

97. (a) Clark, R. H. et al. (2000) 'Nutritional Treatment for Acquired Immunodeficiency Virus-Associated Wasting Using Beta-Hydroxy Beta-Methylbutyrate, Glutamine, and Arginine: A Randomized, Double-Blind, Placebo-Controlled Study', *Journal of Parenteral and Enteral Nutrition*, 24(3), 133–9; (b) Eubanks May, P. et al. (2002) 'Reversal of Cancer-Related Wasting Using Oral Supplementation with a Combination of β-Hydroxy-β-Methylbutyrate, Arginine, and Glutamine', *American Journal of Surgery*, 183(4), 471–9; (c) Rathmacher, J. A. et al. (2004) 'Supplementation with a Combination of Beta-Hydroxy-Beta-Methylbutyrate (HMB), Arginine, and Glutamine Is Safe and Could Improve Hematological Parameters', *Journal of Parenteral and Enteral Nutrition*, 28(2), 65–75.

98. (a) Kreider, R. B. et al. (1999) 'Effects of Calcium Beta-Hydroxy-Beta-Methylbutyrate (HMB) Supplementation During Resistance-Training on Markers of Catabolism, Body Composition and Strength', *International Journal of Sports Medicine*, 20(8), 503–9; (b) Kreider, R. B. et al. (2000) 'Effects of Calcium B-HMB Supplementation During Training on Markers of Catabolism, Body Composition, Strength, and Sprint Performance', *Journal of Exercise Physiology*, 3(4), 48–59; (c) Hoffman, J. R. et al. (2004) 'Effects of Beta-Hydroxy Beta-Methylbutyrate on Power Performance and Indices of Muscle Damage and Stress During High-Intensity Training', *Journal of Strength and Conditioning Research*, 18(4), 747–52; (d) Nunan, D. et al. (2010) 'Exercise-Induced Muscle Damage Is Not Attenuated by Beta-Hydroxy-Beta-Methylbutyrate and Alpha-Ketoisocaproic Acid Supplementation', *Journal of Strength and Conditioning Research*, 24(2), 531–7.

99. Grand View Research (2024) *Energy Drinks Market Size, Share & Trend Analysis Report by Product (Energy Drinks, Energy Shots), by Type (Organic, Conventional), by Packaging, by Distribution Channel, by Region, and Segment Forecasts, 2024–2030*. Available at: https://www.

grandviewresearch.com/industry-analysis/energy-drinks-market (Accessed: 24 July 2024).

100. Hill, A. (2013) 'Dead Marathon Runner Had Now-Banned Stimulant in Water Bottle', *The Guardian*, 30 January. Available at: https://www. theguardian.com/uk/2013/jan/30/dead-marathon-runner-banned-stimulant-dmaa (Accessed: 24 July 2024).

101. US Food & Drug Administration (2018) *DMAA in Products Marked as Dietary Supplements*. Available at: https://www.fda.gov/food/information-select-dietary-supplement-ingredients-and-other-substances/ dmaa-products-marketed-dietary-supplements (Accessed: 25 July 2024).

102. Temple, J. L. et al. (2017) 'The Safety of Ingested Caffeine: A Comprehensive Review', *Frontiers in Psychiatry*, 8, 80.

103. Pollan, M. (2009) *Food Rules: An Eater's Manual*, London: Penguin, p. 103.

Chapter 5: Contemplating Food and Our Wellbeing

1. Course Hero (n.d.) *History of Mental Illness from the Stone Age to the 20th Century*. Available at: https://courses.lumenlearning.com/hvcc-abnormalpsychology/chapter/1-5-prominent-themes-in-abnormal-psychology-throughout-history/ (Accessed: 26 July 2024).

2. World Health Organization (2022) *Mental Disorders*. Available at: https:// www.who.int/news-room/fact-sheets/detail/mental-disorders (Accessed: 29 July 2024).

3. GBD 2017 Global Disease and Injury Incidence and Prevalence Collaborators (2018) 'Global, Regional, and National Incidence, Prevalence, and Years Lived with Disability for 354 Diseases and Injuries for 195 Countries and Territories, 1990–2017: A Systematic Analysis for the Global Burden of Disease Study 2017', *The Lancet*, 392(10159), 1789–858.

4. World Health Organization (2021) *Suicide*. Available at: https://www.who. int/news-room/fact-sheets/detail/suicide (Accessed: 26 July 2024).

5. The number of reported suicides in the UK in 2019 was 6,524 (Mental Health Foundation (n.d.) *Suicide: Statistics*. Available at: https://www. mentalhealth.org.uk/statistics/mental-health-statistics-suicide (Accessed: 26 July 2024)) vs 1,752 reported road deaths (Department for Transport (2020) *Reported Road Casualties in Great Britain: 2019 Annual Report*. Available at: https://assets.publishing.service.gov.uk/government/uploads/ system/uploads/attachment_data/file/922717/reported-road-casualties-annual-report-2019.pdf (Accessed: 26 July 2024)).

6. Henderson, C. et al. (2013) 'Mental Illness Stigma, Help Seeking, and Public Health Programs', *American Journal of Public Health*, 103(5), 777–80.

7. NHS (2023) *Phenylketonuria*. Available at: https://www.nhs.uk/conditions/phenylketonuria/ (Accessed: 29 July 2024).

8. NHS (2023) *Overview: Haemochromatosis*. Available at: https://www.nhs.uk/conditions/haemochromatosis/ (Accessed: 29 July 2024).

9. Stierman, B. et al. (2020) *Special Diets Among Adults: United States, 2015–2018*. National Center for Health Statistics. Available at: https://www.cdc.gov/nchs/products/databriefs/db389.htm (Accessed: 26 July 2024).

10. International Food Information Council (2021) *2021 Food & Health Survey*. Available at: https://foodinsight.org/wp-content/uploads/2021/05/IFIC-2021-Food-and-Health-Survey.May-2021-1.pdf (Accessed: 26 July 2024).

11. Kretsch, M. et al. (1997) 'Cognitive Effects of a Long-Term Weight Reducing Diet', *International Journal of Obesity*, 21, 14–21.

12. Higgs, S. and Spetter, M. S. (2018) 'Cognitive Control of Eating: The Role of Memory in Appetite and Weight Gain', *Current Obesity Reports*, 7, 50–9.

13. (a) Meule, A. (2020) 'The Psychology of Food Cravings: The Role of Food Deprivation', *Current Nutrition Reports*, 9, 251–7; (b) Myers, C. A. et al. (2018) 'Food Cravings and Body Weight: A Conditioning Response', *Current Opinion in Endocrinology, Diabetes and Obesity*, 25(5), 298–302.

14. Gailliot, M. T. and Baumeister, R. F. (2007) 'The Physiology of Willpower: Linking Blood Glucose to Self-Control', *Personality and Social Psychology Review*, 11(4), 303–27.

15. Gailliot, M. T. et al. (2007) 'Self-Control Relies on Glucose as a Limited Energy Source: Willpower Is More Than a Metaphor', *Journal of Personality and Social Psychology*, 92(2), 325–36.

16. (a) Ye, X. et al. (2011) 'Habitual Sugar Intake and Cognitive Function Among Middle-Aged and Older Puerto Ricans Without Diabetes', *British Journal of Nutrition*, 106(9), 1423–32; (b) Barnes, J. N. and Joyner, M. J. (2012) 'Sugar Highs and Lows: The Impact of Diet on Cognitive Function', *Journal of Physiology*, 590(12), 2831; (c) Chong, C. P. et al. (2019) 'Habitual Sugar Intake and Cognitive Impairment Among Multi-Ethnic Malaysian Older Adults', *Clinical Interventions in Aging*, 14, 1331–42.

17. Ginieis, G. et al. (2018) 'The "Sweet" Effect: Comparative Assessments of Dietary Sugars on Cognitive Performance', *Physiology & Behavior*, 184, 242–7.

References

18. Hewlett, P. et al. (2009) 'Grazing, Cognitive Performance and Mood', *Appetite*, 52(1), 245–8.

19. Kanarek, R. (1997) 'Psychological Effects of Snacks and Altered Meal Frequency', *British Journal of Nutrition*, 77(S1), s105–20.

20. Gupta, C. C. et al. (2019) 'Altering Meal Timing to Improve Cognitive Performance During Simulated Nightshifts', *Chronobiology International*, 36(12), 1691–713.

21. (a) Oosterbeek, H. and van der Klaauw, B. (2013) 'Ramadan, Fasting and Educational Outcomes', *Economics of Education Review*, 34, 219–26; (b) Masismadi, N. A. et al. (2017) 'Ramadan Fasting and the Propensity for Learning: Is There a Cause for Concern?', *The Clearing House: A Journal of Educational Strategies, Issues and Ideas*, 90(3), 77–85; (c) Cherif, A. et al. (2016) 'Effects of Intermittent Fasting, Caloric Restriction, and Ramadan Intermittent Fasting on Cognitive Performance at Rest and During Exercise in Adults', *Sports Medicine*, 46, 35–47.

22. Mohd Yasin, W. et al. (2013) 'Does Religious Fasting Affect Cognitive Performance?', *Nutrition & Food Science*, 43(5), 483–9.

23. Benau, E. M. et al. (2021) 'How Does Fasting Affect Cognition? An Updated Systematic Review (2013–2020)', *Current Nutrition Reports*, 10, 376–90.

24. Examples: (a) Ooi, T. C. et al. (2020) 'Intermittent Fasting Enhanced the Cognitive Function in Older Adults with Mild Cognitive Impairment by Inducing Biochemical and Metabolic Changes: A 3-Year Progressive Study', *Nutrients*, 12(9), 2644; (b) Currenti, W. et al. (2021) 'Association Between Time Restricted Feeding and Cognitive Status in Older Italian Adults', *Nutrients*, 13(1), 191; (c) Dias, G. P. et al. (2021) 'Intermittent Fasting Enhances Long-Term Memory Consolidation, Adult Hippocampal Neurogenesis, and Expression of Longevity Gene Klotho', *Molecular Psychiatry*, 26, 6365–79.

25. Horrocks, L. A. and Yeo, Y. K. (1999) 'Health Benefits of Docosahexaenoic Acid (DHA)', *Pharmacological Research*, 40(3), 211–25.

26. (a) Weiser, M. J. et al. (2016) 'Docosahexaenoic Acid and Cognition Throughout the Lifespan', *Nutrients*, 8(2), 99; (b) Cardoso, C. et al. (2016) 'Dietary DHA and Health: Cognitive Function Ageing', *Nutrition Research Reviews*, 29(2), 281–94.

27. Stonehouse, W. et al. (2013) 'DHA Supplementation Improved Both Memory and Reaction Time in Healthy Young Adults: A Randomized Controlled Trial', *American Journal of Clinical Nutrition*, 97(5), 1134–43.

28. Yurko-Mauro, K. et al. (2010) 'Beneficial Effects of Docosahexaenoic Acid on Cognition in Age-Related Cognitive Decline', *Alzheimer's & Dementia*, 6(6), 456–64.

29. Chang, J. P.-C. et al. (2019) 'High-Dose Eicosapentaenoic Acid (EPA) Improves Attention and Vigilance in Children and Adolescents with Attention Deficit Hyperactivity Disorder (ADHD) and Low Endogenous EPA Levels', *Translational Psychiatry*, 9, 303.

30. British Coffee Association (n.d.) *Coffee Consumption*. Available at: https://britishcoffeeassociation.org/coffee-consumption/ (Accessed: 26 July 2024).

31. National Coffee Association of USA (n.d.) *The History of Coffee*. Available at: https://www.ncausa.org/About-Coffee/History-of-Coffee (Accessed: 29 July 2024).

32. Ribeiro, J. A. and Sebastião, A. M. (2010) 'Caffeine and Adenosine', *Journal of Alzheimer's Disease*, 20(S1), s3–15.

33. Coffee & Health (2021) 'Drinking Coffee Helps to Improve Mood as Days Get Shorter', 29 October. Available at: https://www.coffeeandscience.org/health/media-content/news-alerts/drinking-coffee-helps-to-improve-mood-as-days-get-shorter (Accessed: 26 July 2024).

34. Lucas, M. et al. (2011) 'Coffee, Caffeine, and Risk of Depression Among Women', *Archives of Internal Medicine*, 171(17), 1571–8.

35. Addicott, M. A. (2014) 'Caffeine Use Disorder: A Review of the Evidence and Future Implications', *Current Addiction Reports*, 1(3), 186–92.

36. Pramudya, R. C. and Seo, H.-S. (2018) 'Influences of Product Temperature on Emotional Responses to, and Sensory Attributes of, Coffee and Green Tea Beverages', *Frontiers in Psychology*, 8, 2264.

37. Williams, L. E. and Bargh, J. A. (2008) 'Experiencing Physical Warmth Promotes Interpersonal Warmth', *Science*, 322(5901), 606–7.

38. Pham, N. M. et al. (2014) 'Green Tea and Coffee Consumption Is Inversely Associated with Depressive Symptoms in a Japanese Working Population', *Public Health Nutrition*, 17(3), 625–33.

39. Wang, L. et al. (2016) 'Coffee and Caffeine Consumption and Depression: A Meta-Analysis of Observational Studies', *Australian & New Zealand Journal of Psychiatry*, 50(3), 228–42.

40. Ferré, S. (2008) 'An Update on the Mechanisms of the Psychostimulant Effects of Caffeine', *Journal of Neurochemistry*, 105(4), 1067–79.

41. Lassale, C. et al. (2019) 'Healthy Dietary Indices and Risk of Depressive Outcomes: A Systematic Review and Meta-Analysis of Observational Studies', *Molecular Psychiatry*, 24, 965–86.

References

42. Ludwig, D. S. (2002) 'The Glycemic Index: Physiological Mechanisms Relating to Obesity, Diabetes, and Cardiovascular Disease', *JAMA*, 287(18), 2414–23.

43. Salari-Moghaddam, A. et al. (2019) 'Glycemic Index, Glycemic Load, and Depression: A Systematic Review and Meta-Analysis', *European Journal of Clinical Nutrition*, 73, 356–65.

44. McIntyre, R. S. et al. (2010) 'Brain Volume Abnormalities and Neurocognitive Deficits in Diabetes Mellitus: Points of Pathophysiological Commonality with Mood Disorders?', *Advances in Therapy*, 27(2), 63–80.

45. Dinan, T. G. et al. (2015) 'Collective Unconscious: How Gut Microbes Shape Human Behavior', *Journal of Psychiatric Research*, 63, 1–9.

46. (a) Baothman, O. A. et al. (2016) 'The Role of Gut Microbiota in the Development of Obesity and Diabetes', *Lipids in Health and Disease*, 15, 108; (b) Valdes, A. M. et al. (2018) 'Role of the Gut Microbiota in Nutrition and Health', *BMJ*, 361, k2179.

47. Sudo, N. et al. (2004) 'Postnatal Microbial Colonization Programs the Hypothalamic–Pituitary–Adrenal System for Stress Response in Mice', *Journal of Physiology*, 558, 263–75.

48. Phillips, J. G. P. (1910) 'The Treatment of Melancholia by the Lactic Acid Bacillus', *Mental Science*, 56(234), 422–30.

49. Kelly, J. R. et al. (2016) 'Transferring the Blues: Depression-Associated Gut Microbiota Induces Neurobehavioural Changes in the Rat', *Journal of Psychiatric Research*, 82, 109–18.

50. Holobiome (2024) *Holobiome*. Available at: https://holobiome.org/ (Accessed: 29 July 2024).

51. Dinan, T. G. et al. (2013) 'Psychobiotics: A Novel Class of Psychotropic', *Biological Psychiatry*, 74(10), 720–6.

52. (a) Ibid.; (b) Strandwitz, P. (2018) 'Neurotransmitter Modulation by the Gut Microbiota', *Brain Research*, 1693(B), 128–33.

53. Bravo, J. A. et al. (2011) 'Ingestion of Lactobacillus Strain Regulates Emotional Behavior and Central GABA Receptor Expression in a Mouse via the Vagus Nerve', *Proceedings of the National Academy of Sciences of the USA*, 108(38), 16050–5.

54. Camilleri, M. (2009) 'Serotonin in the Gastrointestinal Tract', *Current Opinion in Endocrinology, Diabetes and Obesity*, 16(1), 53–9.

55. Jakobsen, J. C. et al. (2020) 'Should Antidepressants Be Used for Major Depressive Disorder?', *BMJ Evidence-Based Medicine*, 25, 130.

56. McVey Neufeld, K. A. et al. (2019) 'Oral Selective Serotonin Reuptake Inhibitors Activate Vagus Nerve Dependent Gut-Brain Signalling', *Scientific Reports*, 9, 14290.

57. Aaronson, S. T. et al. (2017) 'A 5-Year Observational Study of Patients with Treatment-Resistant Depression Treated with Vagus Nerve Stimulation or Treatment as Usual: Comparison of Response, Remission, and Suicidality', *American Journal of Psychiatry*, 174(7), 640–8.

58. Institute for Quality and Efficiency in Health Care (2020) *Depression: Learn More – How Effective Are Antidepressants?* Available at: https://www.ncbi.nlm.nih.gov/books/NBK361016/ (Accessed: 27 July 2024).

59. World Health Organization (2024) *Alcohol.* Available at: https://www.who.int/news-room/fact-sheets/detail/alcohol (Accessed: 29 July 2024).

60. Albuquerque, N. (2022) *Alcohol Addiction*. UK Addiction Treatment Centres. Available at: https://www.ukat.co.uk/alcohol/ (Accessed: 26 July 2024).

61. Organisation for Economic Co-operation and Development (2021) 'The effect of COVID-19 on alcohol consumption, and policy responses to prevent harmful alcohol consumption', 19 May. Available at: https://www.oecd.org/en/publications/2021/05/the-effect-of-covid-19-on-alcohol-consumption-and-policy-responses-to-prevent-harmful-alcohol-consumption_75c4d184.html (Accessed: 24 July 2024).

62. Public Health England (2021) *Monitoring Alcohol Consumption and Harm during the COVID-19 Pandemic: Summary.* Available at: https://www.gov.uk/government/publications/alcohol-consumption-and-harm-during-the-covid-19-pandemic/monitoring-alcohol-consumption-and-harm-during-the-covid-19-pandemic-summary (Accessed: 26 July 2024).

63. Banerjee, N. (2014) 'Neurotransmitters in Alcoholism: A Review of Neurobiological and Genetic Studies', *Indian Journal of Human Genetics*, 20(1), 20–31.

64. Ibid.

65. Mitchell, J. M. et al. (2012) 'Alcohol Consumption Induces Endogenous Opioid Release in the Human Orbitofrontal Cortex and Nucleus Accumbens', *Science Translational Medicine*, 4(116), 116ra6.

66. Juárez, J. and Molina-Martínez, L. M. (2019) 'Chapter 45 – Opioid System and Alcohol Consumption', in: V. R. Preedy (ed.), *Neuroscience of Alcohol*, Cambridge, MA: Academic Press, 435–42.

67. Banerjee, 'Neurotransmitters in Alcoholism'.

References

68. Blaine, S. K. et al. (2016) 'Alcohol Effects on Stress Pathways: Impact on Craving and Relapse Risk', *Canadian Journal of Psychiatry*, 61(3), 145–53.

69. Kokavec, A. (2012) 'Salivary or Serum Cortisol: Possible Implications for Alcohol Research', in S. M. Ostojic (ed.) *Steroids: From Physiology to Clinical Medicine*, IntechOpen. Available at: https://www.intechopen.com/chapters/38178 (Accessed: 27 July 2024).

70. Gilpin, N. W. and Koob, G. F. (2008) 'Neurobiology of Alcohol Dependence: Focus on Motivational Mechanisms', *Alcohol Research & Health*, 31(3), 185–95.

71. Fleming, A. (2019) '"Hangxiety": Why Alcohol Gives You a Hangover and Anxiety', *The Guardian*, 27 January. Available at: https://www.theguardian.com/lifeandstyle/2019/jan/27/hangxiety-why-alcohol-gives-you-a-hangover-and-anxiety (Accessed: 26 July 2024).

72. (a) Thayer, J. F. et al. (2006) 'Alcohol Use, Urinary Cortisol, and Heart Rate Variability in Apparently Healthy Men: Evidence for Impaired Inhibitory Control of the HPA Axis in Heavy Drinkers', *International Journal of Psychophysiology*, 59(3), 244–50; (b) Dai, X. et al. (2007) 'Response of the HPA-Axis to Alcohol and Stress as a Function of Alcohol Dependence and Family History of Alcoholism', *Psychoneuroendocrinology*, 32(3), 293–305.

73. Badrick, E. et al. (2008) 'The Relationship between Alcohol Consumption and Cortisol Secretion in an Aging Cohort', *Journal of Clinical Endocrinology & Metabolism*, 93(3), 750–7.

74. (a) Rose, A. K. et al. (2010) 'The Importance of Glucocorticoids in Alcohol Dependence and Neurotoxicity', *Alcoholism: Clinical and Experimental Research*, 34, 2011–18; (b) Blaine, S. K. et al. (2019) 'Craving, Cortisol and Behavioral Alcohol Motivation Responses to Stress and Alcohol Cue Contexts and Discrete Cues in Binge and Non-Binge Drinkers', *Addiction Biology*, 24, 1096–108.

75. Gilpin and Koob, 'Neurobiology of Alcohol Dependence'.

76. Jackson, S. E. et al. (2019) 'Is There a Relationship Between Chocolate Consumption and Symptoms of Depression? A Cross-Sectional Survey of 13,626 US Adults', *Depression and Anxiety*, 36, 987–95.

77. Scholey, A. and Owen, L. (2013) 'Effects of Chocolate on Cognitive Function and Mood: A Systematic Review', *Nutrition Reviews*, 71(10), 665–81.

Chapter 6: Contemplating the Sustainability of Our Food

1. United Nations Development Programme and University of Oxford (2021) *The Peoples' Climate Vote*. Available at: https://www.undp.org/content/undp/en/home/librarypage/climate-and-disaster-resilience-/The-Peoples-Climate-Vote-Results.html (Accessed: 26 July 2024).

2. Powell, J. (2017) 'Scientists Reach 100% Consensus on Anthropogenic Global Warming', *Bulletin of Science, Technology & Society*, 37(4), 183–4.

3. Poore, J. and Nemecek, T. (2018) 'Reducing Food's Environmental Impacts Through Producers and Consumers', *Science*, 360(6392), 987–92.

4. Gates, B. (2021) *How to Avoid a Climate Disaster: The Solutions We Have and the Breakthroughs We Need*, London: Allen Lane.

5. Food and Agriculture Organization of the United Nations (2006) *Livestock's Long Shadow: Environmental Issues and Options*, Rome: Food and Agriculture Organization of the United Nations.

6. Ritchie, H. et al. (2024) 'Greenhouse Gas Emissions', *Our World in Data*. Available at: https://ourworldindata.org/greenhouse-gas-emissions (Accessed: 25 July 2024).

7. United Nations Environment Programme (2019) *Emissions Gap Report 2019*, Nairobi: United Nations Environment Programme.

8. Urban, M. C. (2015) 'Accelerating Extinction Risk from Climate Change', *Science*, 348(6234), 571–3.

9. United Nations Treaty Collection (2015) *Chapter XXVII: Environment. 7.d Paris Agreement*. Available at: https://treaties.un.org/pages/ViewDetails.aspx?src=TREATY&mtdsg_no=XXVII-7-d&chapter=27&clang=_en (Accessed: 26 July 2024).

10. Liu, P. R. and Raftery, A. E. (2021) 'Country-Based Rate of Emissions Reductions Should Increase by 80% Beyond Nationally Determined Contributions to Meet the 2°C Target', *Communications Earth & Environment*, 2, 29.

11. Figures from Worldometer (2023) *World Population Clock*. Available at: https://www.Worldometers.info/world-population (Accessed: 26 July 2024).

12. Poore and Nemecek, 'Reducing Food's Environmental Impacts Through Producers and Consumers'.

13. Intergovernmental Panel on Climate Change (2019) *Climate Change and Land: An IPCC Special Report on Climate Change, Desertification, Land Degradation, Sustainable Land Management, Food Security, and*

Greenhouse Gas Fluxes in Terrestrial Ecosystem. Available at: https://www.ipcc.ch/srccl/ (Accessed: 26 July 2024).

14. United Nations Environment Programme and Climate and Clean Air Coalition (2021) *Global Methane Assessment: Benefits and Costs of Mitigating Methane Emissions*, Nairobi: United Nations Environment Programme.

15. Ibid.

16. *The Economist* (2021) 'Those Who Worry About CO_2 Should Worry About Methane, Too', 3 April. Available at: https://www.economist.com/science-and-technology/2021/04/03/those-who-worry-about-co2-should-worry-about-methane-too (Accessed: 26 July 2024).

17. United Nations Environment Programme and Climate and Clean Air Coalition, *Global Methane Assessment*.

18. Ibid.

19. Data from Ritchie, H. et al. (2023) 'Meat and Dairy Production', *Our World in Data*. Available at: https://ourworldindata.org/meat-production (Accessed: 25 July 2024).

20. (a) Statista (2023) *Cow Milk Production Worldwide from 2015 to 2022*. Available at: https://www.statista.com/statistics/263952/production-of-milk-worldwide (Accessed: 25 July 2024); (b) Statista (2023) *Global Egg Production from 1990 to 2021 (in 1,000 Metric Tons)*. Available at: https://www.statista.com/statistics/263972/egg-production-worldwide-since-1990/ (Accessed: 25 July 2024).

21. Ritchie et al., *Meat and Dairy Production*.

22. Intergovernmental Science-Policy Platform (2019) *The Global Assessment Report on Biodiversity and Ecosystem Services: Summary for Policymakers*. Available at: https://ipbes.net/sites/default/files/inline/files/ipbes_global_assessment_report_summary_for_policymakers.pdf (Accessed: 26 July 2024).

23. Ibid.

24. Griscom, B. W. et al. (2019) 'We Need Both Natural and Energy Solutions to Stabilize Our Climate', *Global Change Biology*, 25(6), 188–990.

25. Vermeulen, S. J. et al. (2012) 'Climate Change and Food Systems', *Annual Review of Environment and Resources*, 37, 195–222.

26. Poore and Nemecek, 'Reducing Food's Environmental Impacts Through Producers and Consumers'.

27. Tilman, D. et al. (2017) 'Future Threats to Biodiversity and Pathways to Their Prevention', *Nature*, 546, 73–81.

28. National Food Strategy (2021) *The National Food Strategy: The Plan*. Available at: https://www.nationalfoodstrategy.org/ (Accessed: 27 July 2024), p. 15.
29. Montgomery, D. R. et al. (2022) 'Soil Health and Nutrient Density: Preliminary Comparison of Regenerative and Conventional Farming', *PeerJ*, 10, e12848.
30. Holmes, H. (2021) 'Regenerative Agriculture: Why PepsiCo, McDonald's and Waitrose Are Jumping on Regen Ag', *The Grocer*, 23 August. Available at: https://www.thegrocer.co.uk/sustainability-and-environment/ regenerative-agriculture-why-pepsico-mcdonalds-and-waitrose-are-jumping-on-regen-ag/659119.article (Accessed: 26 July 2024).
31. Ritchie, H. and Roser, M. (2024) 'Water Use and Stress', *Our World in Data*, February. Available at: https://ourworldindata.org/water-use-stress (Accessed: 29 July 2024).
32. European Environment Agency (2023) *Water Use and Environmental Pressures*. Available at: https://www.eea.europa.eu/themes/water/ european-waters/water-use-and-environmental-pressures (Accessed: 26 July 2024).
33. Khokhar, T. (2017) 'Chart: Globally, 70% of Freshwater Is Used for Agriculture', *World Bank Blogs*, 22 March. Available at: https://blogs. worldbank.org/opendata/chart-globally-70-freshwater-used-agriculture (Accessed: 26 July 2024).
34. Mekonnen, M. M. and Hoekstra, A. Y. (2012) 'A Global Assessment of the Water Footprint of Farm Animal Products', *Ecosystems*, 15(3), 401–15.
35. Ibid.
36. Ibid.
37. Ertug Ercin, A. et al. (2012) 'The Water Footprint of Soy Milk and Soy Burger and Equivalent Animal Products', *Ecological Indicators*, 18, 392–402.
38. Chapagain, A. and Hoekstra, A. (2008) 'The Global Component of Freshwater Demand and Supply: An Assessment of Virtual Water Flows Between Nations as a Result of Trade in Agricultural and Industrial Products', *Water International*, 33, 19–32.
39. Intergovernmental Panel on Climate Change, *Climate Change and Land*.
40. United Nations Environment Programme and Climate and Clean Air Coalition, *Global Methane Assessment*.
41. Food and Agriculture Organization of the United Nations (n.d.) *Technical Platform on the Measurement and Reduction of Food Loss and Waste: Food*

References

Loss. Available at: https://www.fao.org/platform-food-loss-waste/food-loss/introduction/en (Accessed: 29 July 2024).

42. Food and Agriculture Organization of the United Nations (n.d.) *Technical Platform on the Measurement and Reduction of Food Loss and Waste: Food Waste*. Available at: https://www.fao.org/platform-food-loss-waste/food-waste/introduction/en (Accessed: 29 July 2024).

43. Food and Agriculture Organization of the United Nations (2021) 'The Scourge of Food Loss and Waste Needs to Be Urgently Tackled to Achieve the World's 2030 Target', 29 September. Available at: https://wrap.ngo/resources/report/courtauld-commitment-2025-annual-report-2020 (Accessed: 29 July 2024).

44. Waste and Resources Action Programme (2020) *Courtauld Commitment 2025: Annual Report 2020*. Available at: https://wrap.org.uk/sites/default/files/2021-01/The-Courtauld-Commitment-2025-Annual_Report-2020.pdf (Accessed: 29 July 2024).

45. Waste Managed (2024) 'Food Waste – 2024 Facts & Statistics'. Available at: https://www.wastemanaged.co.uk/our-news/food-waste/food-waste-facts-statistics/ (Accessed: 29 July 2024).

46. Dwyer, O. (2023) 'Food Waste Makes Up "Half" of Global Food System Emissions', *Carbon Brief*, 13 March. Available at: https://www.carbonbrief.org/food-waste-makes-up-half-of-global-food-system-emissions/ (Accessed: 29 July 2024).

47. Mekonnen, M. M. and Hoekstra, A. Y. (2011) 'The Green, Blue and Grey Water Footprint of Crops and Derived Crop Products', *Hydrology and Earth System Sciences*, 15, 1577–600.

48. Kummu, M. et al. (2012) 'Lost Food, Wasted Resources: Global Food Supply Chain Losses and Their Impacts on Freshwater, Cropland, and Fertiliser Use', *Science of the Total Environment*, 438, 477–89.

49. Wasteless (2024) *Wasteless*. Available at: www.wasteless.com (Accessed: 29 July 2024).

50. Ritchie, H. and Roser, M. (2021) 'Fish and Overfishing', *Our World in Data*. Available at: https://ourworldindata.org/seafood-production (Accessed: 26 July 2024).

51. Ibid.

52. Ibid.

53. Krauss, L. M. (2021) *The Physics of Climate Change*, London: Head of Zeus, p. 148.

54. Food and Agriculture Organization of the United Nations (n.d.) *Appendix 2: Facsimile from the Book of Fan Lai*. Available at: https://www.fao.org/3/ag158e/AG158E04.htm#app2 (Accessed: 26 July 2024).

55. Food and Agriculture Organization of the United Nations (2020) *The State of World Fisheries and Aquaculture: Sustainability in Action*. Available at: https://www.fao.org/3/ca9229en/ca9229en.pdf (Accessed: 26 July 2024).

56. Gephart, J. A. et al. (2021) 'Environmental Performance of Blue Foods', *Nature*, 597, 360–5.

57. Paddison, L. (2021) 'Fish Farmers Grapple with Sustainability Challenge', *Financial Times*, 1 November. Available at: https://www.ft.com/content/5a42b73d-0ac7-4403-89ef-446f14ee61c1 (Accessed: 26 July 2024).

58. Food and Agriculture Organization of the United Nations, *The State of World Fisheries and Aquaculture*.

59. Ibid.

60. DeWeerdt, S. (2020) 'Can Aquaculture Overcome Its Sustainability Challenges?', *Nature*, 18 December. Available at: https://www.nature.com/articles/d41586-020-03446-3 (Accessed: 26 July 2024).

61. Willer, D. F. et al. (2022) 'Maximising Sustainable Nutrient Production from Coupled Fisheries-Aquaculture Systems', *PLOS Sustainability and Transformation*, 1(3), e000000.

62. Fishfarmingexpert (2021) '"All Fish Lost" in Blaze at Atlantic Sapphire Denmark', 16 December. Available at: https://www.fishfarmingexpert.com/article/atlantic-sapphire-loses-all-in-fish-in-blaze-at-danish-site/ (Accessed: 26 July 2024).

63. Liu, Y. et al. (2016) 'Comparative Economic Performance and Carbon Footprint of Two Farming Models for Producing Atlantic Salmon (Salmo Salar): Land-Based Closed Containment System in Freshwater and Open Net Pen in Seawater', *Aquacultural Engineering*, 71, 1–12.

64. Food and Agriculture Organization of the United Nations, *The State of World Fisheries and Aquaculture*.

65. Sandström, V. et al. (2018) 'The Role of Trade in the Greenhouse Gas Footprints of EU Diets', *Global Food Security*, 19, 48–55.

66. (a) Public Health England (2018) *National Diet and Nutrition Survey: Results from Years 7 and 8 (Combined) of the Rolling Programme (2014/2015 to 2015/2016)*. Available at: https://assets.publishing.service.gov.uk/government/uploads/system/uploads/attachment_data/file/699241/NDNS_results_years_7_and_8.pdf (Accessed: 26 July 2024); (b) US Department of Agriculture (2023) *Food Availability (Per Capita) Data*

System: Loss-Adjusted Food Availability. Available at: https://www.ers.usda.gov/data-products/food-availability-per-capita-data-system/loss-adjusted-food-availability-documentation/ (Accessed: 26 July 2024).

67. Willett, W. et al. (2019) 'Food in the Anthropocene: The EAT–Lancet Commission on Healthy Diets from Sustainable Food Systems', *The Lancet*, 393(10170), 447–92.

68. Tilman, D. and Clark, M. (2014) 'Global Diets Link Environmental Sustainability and Human Health', *Nature*, 515, 518–22.

69. (a) Aleksandrowicz, L. et al. (2016) 'The Impacts of Dietary Change on Greenhouse Gas Emissions, Land Use, Water Use, and Health: A Systematic Review', *PLoS One*, 11(11), e0165797; (b) Peters, C. J. et al. (2016) 'Carrying Capacity of U.S. Agricultural Land: Ten Diet Scenarios', *Elementa: Science of the Anthropocene*, 4, 000116.

70. Vanham, D. et al. (2018) 'The Water Footprint of Different Diets Within European Sub-National Geographical Entities', *Nature Sustainability*, 1, 518–25.

71. Poore and Nemecek, 'Reducing Food's Environmental Impacts Through Producers and Consumers'.

72. Phalan, B. (2018) 'What Have We Learned from the Land Sparing-Sharing Model?', *Sustainability*, 10(6), 1760.

73. As defined in Carrera-Bastos, P. et al. (2011) 'The Western Diet and Lifestyle and Diseases of Civilization', *Research Reports in Clinical Cardiology*, 2, 15–35.

74. Food and Agriculture Organization of the United Nations, *Livestock's Long Shadow*.

75. United Nations Development Programme and University of Oxford, *The Peoples' Climate Vote*.

76. Huel (2020) *Huel Sustainable Nutrition Report 2020*. Available at: https://view.publitas.com/huel/huel-sustainable-nutrition-report-2020/page/1 (Accessed: 26 July 2024).

77. (a) Scarborough, P. et al. (2014) 'Dietary Greenhouse Gas Emissions of Meat-Eaters, Fish-Eaters, Vegetarians and Vegans in the UK', *Climatic Change*, 125, 179–92; (b) Whitton, C. et al. (2011) 'National Diet and Nutrition Survey: UK Food Consumption and Nutrient Intakes from the First Year of the Rolling Programme and Comparisons with Previous Surveys', *British Journal of Nutrition*, 106(12), 1899–914.

78. (a) Public Health England, *National Diet and Nutrition Survey*; (b) US Department of Agriculture, *Food Availability (Per Capita) Data System*.

Chapter 7: Contemplating the Ethics of What We Eat

1. Berger (1977), cited by Berger, J. (2009) *Why Look at Animals?*, London: Penguin.
2. Pollan, M. (2006) *The Omnivore's Dilemma: The Search for a Perfect Meal in a Fast-Food World*, London: Penguin, p. 307.
3. Scientific Group of the UN Food Systems Summit (2021) *The True Cost and True Price of Food*. Available at: https://sc-fss2021.org/wp-content/uploads/2021/06/UNFSS_true_cost_of_food.pdf (Accessed: 27 July 2024).
4. Boyd, D. R. (2021) 'Human Rights Could Address the Health and Environmental Costs of Food Production', *UBC Sustainability*, 23 September. Available at: https://sustain.ubc.ca/stories/human-rights-could-address-health-and-environmental-costs-food-production (Accessed: 27 July 2024).
5. International Labour Organization (n.d.) *Forced Labour, Modern Slavery and Trafficking in Persons*. Available at: https://www.ilo.org/global/topics/forced-labour-modern-slavery-and-traffficking-in-persons (Accessed: 27 July 2024).
6. Askew, K. (2021) 'Why Human Rights Must Be at the Heart of Food Industry Climate Action', *FoodNavigator Europe*, 27 October. Available at: https://www.foodnavigator.com/Article/2021/10/25/Why-human-rights-must-be-at-the-heart-of-food-industry-climate-action (Accessed: 27 July 2024).
7. (a) Fairtrade Foundation (2024) *Fairtrade*. Available at: www.fairtrade.org.uk (Accessed: 29 July 2024); (b) Rainforest Alliance (2024) *Rainforest Alliance*. Available at: www.rainforest-alliance.org (Accessed: 29 July 2024).
8. Jékely, G. (2007) 'Origin of Phagotrophic Eukaryotes as Social Cheaters in Microbial Biofilms', *Biology Direct*, 2, 3.
9. Plotnick, R. et al. (2010) 'Information Landscapes and Sensory Ecology of the Cambrian Radiation', *Paleobiology*, 36(2), 303–17.
10. Kaplan, M. (2012) 'Primates Were Always Tree-Dwellers', *Nature*. Available at: https://www.nature.com/articles/nature.2012.11423 (Accessed: 27 July 2024).
11. Pontzer, H. (2012) 'Overview of Hominin Evolution', *Nature Education Knowledge*, 3(10), 8.
12. Ibid.
13. Roebroeks, W. and Villa, P. (2011) 'On the Earliest Evidence for Habitual

References

Use of Fire in Europe', *Proceedings of the National Academy of Sciences*, 108(13), 5209–14.

14. Ritchie, H. et al. (2023) 'Meat and Dairy Production', *Our World in Data*. Available at: https://ourworldindata.org/meat-production (Accessed: 25 July 2024).

15. Godfray, H. C. J. et al. (2018) 'Meat Consumption, Health, and the Environment', *Science*, 361(6399), eaam5324.

16. Ibid.

17. (a) Berridge, K. C. et al. (2010) 'The Tempted Brain Eats: Pleasure and Desire Circuits in Obesity and Eating Disorders', *Brain Research*, 1350, 43–64; (b) Drewnowski, A. et al. (2012) 'Sweetness and Food Preference', *Journal of Nutrition*, 142(6), 1142–8s.

18. (a) Marteau, T. M. (2017) 'Towards Environmentally Sustainable Human Behaviour: Targeting Non-Conscious and Conscious Processes for Effective and Acceptable Policies', *Philosophical Transactions. Series A: Mathematical, Physical, and Engineering Sciences*, 375(2095), 20160371; (b) Godfray et al., 'Meat Consumption, Health, and the Environment'.

19. Piazza, J. et al. (2015) 'Rationalizing Meat Consumption. The 4Ns', *Appetite*, 91, 114–28.

20. Godfray et al., 'Meat Consumption, Health, and the Environment'.

21. Singer, P. (1975) *Animal Liberation* (2015 edn), London: The Bodley Head, p. 137.

22. In ibid., Yuval Noah Harari, 'Introduction', p. xiv.

23. Ibid, pp. xi–xii.

24. Ibid., p. x.

25. Ibid., p. 109.

26. Ibid., p. 100.

27. Bridgeman, L. (2021) 'What Are Broiler Chickens and How Long Do They Live?', *The Humane League*, 11 February. Available at: https://thehumaneleague.org.uk/article/broiler-chickens (Accessed: 27 July 2024).

28. Singer, *Animal Liberation*, p. 118.

29. Nanaji Deshmukh Veterinary Science University, Jabalpur (n.d.) *Feeding of Broiler*. Available at: http://www.ndvsu.org/images/StudyMaterials/Nutrition/Broiler-nutrition.pdf (Accessed: 27 July 2024).

30. Singer, *Animal Liberation*, p. 99.

31. Royal Society for the Prevention of Cruelty to Animals Australia (2019) *Can Pigs Be Kept as Pets?* Available at: https://kb.rspca.org.au/knowledge-base/can-pigs-be-kept-as-pets/ (Accessed: 27 July 2024).

32. Singer, *Animal Liberation*, pp. 120–9.

33. Ibid., pp. 136–41.

34. Ibid., pp. 129–36.

35. Meat Your Future (2016) 'Temple Grandin and Confusing Ethics About Animals', 29 August. Available at: https://meatyourfuture.com/2016/08/temple-grandin/ (Accessed: 27 July 2024).

36. Yale Center for Business and the Environment (2016) 'Disrupting Meat', 12 October. Available at: https://cbey.yale.edu/our-stories/disrupting-meat (Accessed: 27 July 2024).

37. Singer, *Animal Liberation*, p. 137.

38. Ryder, R. D. (2000) *Animal Revolution: Changing Attitudes Towards Speciesism*, Oxford: Berg Publishers.

39. Bentham, J. (1780) *An Introduction to the Principles of Morals and Legislation*, London: T. Payne and Sons, Chapter 17.

40. Clark, D. (2016) 'Nineteenth-Century Doctors and Care of the Dying', in: *To Comfort Always: A History of Palliative Medicine Since the Nineteenth Century*, Oxford: Oxford University Press, 1-32. Available at: https://oxfordmedicine.com/view/10.1093/med/9780199674282.001.0001/med-9780199674282-chapter-1 (Accessed: 27 July 2024).

41. Ison, S. H. et al. (2016) 'A Review of Pain Assessment in Pigs', *Frontiers in Veterinary Science*, 3, 108.

42. Gentle, M. J. (2011) 'Pain Issues in Poultry', *Applied Animal Behaviour Science*, 135(3), 252–8.

43. Washburn, S. P. et al. (2002) 'Reproduction, Mastitis, and Body Condition of Seasonally Calved Holstein and Jersey Cows in Confinement or Pasture Systems', *Journal of Dairy Science*, 85(1), 105–11.

44. *Animal Welfare Act 2006*, c. 45. Available at: https://www.legislation.gov.uk/ukpga/2006/45/contents (Accessed: 27 July 2024).

45. Shriver, A. (2010) 'Not Grass-Fed, but at Least Pain-Free', *The New York Times*, 18 February. Available at: https://www.nytimes.com/2010/02/19/opinion/19shriver.html (Accessed: 27 July 2024).

46. Nieuwhof, G. J. and Bishop, S. C. (2005) 'Costs of the Major Endemic Diseases of Sheep in Great Britain and the Potential Benefits of Reduction in Disease Impact', *Animal Science*, 81(1), 23–9.

47. Wassink, G. J. et al. (2010) 'A Within Farm Clinical Trial to Compare Two Treatments (Parenteral Antibacterials and Hoof Trimming) for Sheep Lame with Footrot', *Preventive Veterinary Medicine*, 96(1–2), 93–103.

References

48. Compassion in World Farming (n.d.) *Rethink Fish*. Available at: https://www.ciwf.org.uk/our-campaigns/rethink-fish/ (Accessed: 27 July 2024).

49. Hansen, L. P. et al. (1993) 'Oceanic Migration in Homing Atlantic Salmon', *Animal Behaviour*, 45(5), 927–41.

50. Brown, C. (2015) 'Fish Intelligence, Sentience and Ethics', *Animal Cognition*, 18, 1–17.

51. (a) Yue, S. et al. (2004) 'Investigating Fear in Domestic Rainbow Trout, Oncorhynchus Mykiss, Using an Avoidance Learning Task', *Applied Animal Behaviour Science*, 87(3–4), 343–54; (b) Dunlop, R. et al. (2006) 'Avoidance Learning in Goldfish (Carassius Auratus) and Trout (Oncorhynchus Mykiss) and Implications for Pain Perception', *Applied Animal Behaviour Science*, 97(2–4), 255–71; (c) Pittman, J. T. et al. (2013) 'iPhone® Applications as Versatile Video Tracking Tools to Analyze Behavior in Zebrafish (Danio Rerio)', *Pharmacology Biochemistry and Behavior*, 106, 137–42.

52. Key, B. (2016) 'Why Fish Do Not Feel Pain', *Animal Sentience*, 3(1).

53. Compassion in World Farming, *Rethink Fish*.

54. Ibid.

55. Changing Markets Foundation (n.d.) *Fishing the Feed*. Available at: https://changingmarkets.org/portfolio/fishing-the-feed/ (Accessed: 27 July 2024).

56. Food and Agriculture Organization of the United Nations (2016) *The State of World Fisheries and Aquaculture: Contributing to Food Security and Nutrition for All*. Available at: https://www.fao.org/3/i5555e/i5555e.pdf (Accessed: 27 July 2024).

57. Marine Stewardship Council (2024). *Marine Stewardship Council*. Available at: www.msc.org/uk (Accessed: 27 July 2024).

58. Compassion in World Farming, *Rethink Fish*.

59. Martin, M. J. et al. (2015) 'Antibiotics Overuse in Animal Agriculture: A Call to Action for Health Care Providers', *American Journal of Public Health*, 105(12), 2409–10.

60. US Food & Drug Administration (2020) 'All About BSE (Mad Cow Disease)'. Available at: https://www.fda.gov/animal-veterinary/animal-health-literacy/all-about-bse-mad-cow-disease (Accessed: 27 July 2024).

61. BBC News (2018) '"Mad Cow Disease": What Is BSE?', 18 October. Available at: https://www.bbc.co.uk/news/uk-45906585 (Accessed: 27 July 2024).

62. Department for Environment, Food & Rural Affairs (2024) *Animal Health and Welfare Pathway*. Available at: https://www.gov.uk/government/

publications/animal-health-and-welfare-pathway/animal-health-and-welfare-pathway (Accessed: 29 July 2024).

63. Hausman, D. M. and Welch, B. (2010) 'Debate: To Nudge or Not to Nudge*', *Journal of Political Philosophy*, 18, 123–36.

64. Van Niekerk, T. (2024) 'Vegetarian Statistics – Surprising Facts & Data in 2024', *World Animal Foundation*, 1 June. Available at: https://worldanimal foundation.org/advocate/vegetarian-statistics/ (Accessed: 29 August 2024).

65. Paslakis, G. et al. (2020) 'Prevalence and Psychopathology of Vegetarians and Vegans – Results from a Representative Survey in Germany', *Scientific Reports*, 10, 6840.

66. (a) Muñoz, M. et al. (2009) 'An Update on Iron Physiology', *World Journal of Gastroenterology*, 15(37), 4617–26; (b) Hurrell, R. and Egli, I. (2010) 'Iron Bioavailability and Dietary Reference Values', *American Journal of Clinical Nutrition*, 91(5), 1461–7s.

67. Siegenberg, D. et al. (1991) 'Ascorbic Acid Prevents the Dose-Dependent Inhibitory Effects of Polyphenols and Phytates on Nonheme-Iron Absorption', *American Journal of Clinical Nutrition*, 53(2), 537–41.

68. Browning, H. (2018) 'Dairy Farming, Is Organic Really Different?', *Soil Association*, 28 August. Available at: https://www.soilassociation.org/news/2018/august/28/dairy-farming-is-organic-really-different/ (Accessed: 27 July 2024).

69. Smithers, R. (2019) 'Grubs Up: A Third of Britons Think We'll Be Eating Insects by 2029', *The Guardian*, 2 September. Available at: https://www.theguardian.com/food/2019/sep/02/grubs-up-a-third-of-britons-think-well-be-eating-insects-by-2029 (Accessed: 27 July 2024).

70. Smithers, R. (2018) 'Bug Grub: Sainsbury's to Stock Edible Insects on Shelves in a UK First', *The Guardian*, 17 November. Available at: https://www.theguardian.com/business/2018/nov/17/bug-grub-sainsburys-to-stock-edible-insects-on-shelves-in-a-uk-first (Accessed: 27 July 2024).

71. https://www.statista.com/statistics/882321/edible-insects-market-size-global/ (Accessed: 27 July 2024).

72. Good Food Institute (n.d.) *Plant-Based and Cultivated Meat Innovation | GFI*. Available at: https://gfi.org/ (Accessed: 27 July 2024).

Chapter 8: Contemplating Food and Togetherness

1. Oxford Economics and the National Centre for Social Research (2018) *The Sainsbury's Living Well Index*. Available at: https://www.about.sainsburys.

co.uk/~/media/Files/S/Sainsburys/living-well-index/sainsburys-living-well-index-sep-2018.pdf (Accessed: 28 July 2024).

2. Discussed in: (a) Pollan, M. (2008) *In Defence of Food: The Myth of Nutrition and the Pleasures of Eating*, London: Penguin; (b) Zaraska, M. (2020) *Growing Young: How Friendship, Optimism and Kindness Can Help You Live to 100*, London: Robinson.

3. CulinaryLore (2012) 'Origin of the Phrase "Square Meal"', 10 September. Available at: https://culinarylore.com/food-history:origin-of-phrase-a-square-meal/ (Accessed: 28 July 2024).

4. Online Etymology Dictionary (n.d.) *Breakfast (n.)*. Available at: https://www.etymonline.com/search?q=breakfast (Accessed: 28 July 2024).

5. McMillan, S. (2001) 'What Time Is Dinner?', *History Magazine*, October/November. Available at: https://web.archive.org/web/20171208184059/http://www.history-magazine.com/dinner2.html (Accessed: 7 August 2024).

6. Ibid.

7. Dunsworth, H. M. (2010) 'Origin of the Genus *Homo*', *Evolution: Education and Outreach*, 3, 353–66.

8. Guyenet, S. J. (2017) *The Hungry Brain: Outsmarting the Instincts That Make Us Overeat*, London: Vermilion, pp. 87–94.

9. Jenkinson, A. (2020) *Why We Eat (Too Much): The New Science of Appetite*, London: Penguin, pp. 36–7.

10. Hammons, A. J. and Robart, R. (2021) 'Family Food Environment during the COVID-19 Pandemic: A Qualitative Study', *Children*, 8(5), 354.

11. (a) Berge, J. M. (2009) 'A Review of Familial Correlates of Child and Adolescent Obesity: What Has the 21st Century Taught Us So Far?' *International Journal of Adolescent Medicine and Health*, 21(4), 457–83; (b) Hammons, A. J. and Fiese, B. H. (2011) 'Is Frequency of Shared Family Meals Related to the Nutritional Health of Children and Adolescents?' *Pediatrics*, 127(6), e1565–e1574.

12. (a) Eisenberg, M. E., Olson, R. E., Neumark-Sztainer, D., Story, M. and Bearinger, L. H. (2004) 'Correlations between Family Meals and Psychosocial Well-Being among Adolescents', *Archives of Pediatrics & Adolescent Medicine*, 158(8), 792–6; (b) Elgar, F. J., Craig, W. and Trites, S. J. (2013) 'Family Dinners, Communication, and Mental Health in Canadian Adolescents', *Journal of Adolescent Health*, 52(4), 433–8.

13. Zaraska, *Growing Young*, pp. 94–6.

14. (a) Wellen, K. E. et al. (2003) 'Obesity-Induced Inflammatory Changes in Adipose Tissue', *Journal of Clinical Investigation*, 112(12), 1785–8; (b) Dantzer, R. (2012) 'Depression and Inflammation: An Intricate Relationship', *Biological Psychiatry*, 71(1), 4–5; (c) Dantzer, R. et al. (2008) 'From Inflammation to Sickness and Depression: When the Immune System Subjugates the Brain', *Nature Reviews. Neuroscience*, 9(1), 46–56; (d) Parkitny, L. et al. (2013) 'Inflammation in Complex Regional Pain Syndrome: A Systematic Review and Meta-Analysis', *Neurology*, 80(1), 106–17; (e) Minihane, A. M. et al. (2015) 'Low-Grade Inflammation, Diet Composition and Health: Current Research Evidence and Its Translation', *British Journal of Nutrition*, 114(7), 999–1012.

15. Health and Safety Executive (2023) *Work-Related Stress, Depression or Anxiety Statistics in Great Britain, 2023*. Available at: https://www. littlegreenbutton.com/wp-content/uploads/2023/11/Work-related-stress-depression-or-anxiety-statistics-in-Great-Britain-2023.pdf (Accessed: 29 July 2024).

16. Yau, Y. H. and Potenza, M. N. (2013) 'Stress and Eating Behaviors', *Minerva Endocrinology*, 38(3), 255–67.

17. Pruessner, J. C. et al. (2004) 'Dopamine Release in Response to a Psychological Stress in Humans and Its Relationship to Early Life Maternal Care: A Positron Emission Tomography Study Using Raclopride', *Journal of Neuroscience*, 24(11), 2825–31.

18. (a) Wittig, R. et al. (2016) 'Social Support Reduces Stress Hormone Levels in Wild Chimpanzees Across Stressful Events and Everyday Affiliations', *Nature Communications*, 7, 13361; (b) Stocker, M. et al. (2020) 'Cooperation with Closely Bonded Individuals Reduces Cortisol Levels in Long-Tailed Macaques', *Royal Society Open Science*, 7(5), 191056.

19. Stocker, 'Cooperation with Closely Bonded Individuals Reduces Cortisol Levels in Long-Tailed Macaques'.

20. De Castro, J. M. (1994) 'Family and Friends Produce Greater Social Facilitation of Food Intake than Other Companions', *Physiology & Behavior*, 56(3), 445–5.

21. Raff, K. (2011) 'The Roman Banquet', *The Met*, October. Available at: https://www.metmuseum.org/toah/hd/banq/hd_banq.htm (Accessed: 28 July 2024).

22. Dunbar, R. I. M. (2017) 'Breaking Bread: The Functions of Social Eating', *Adaptive Human Behavior and Physiology*, 3, 198–211.

References

23. Tarr, B. et al. (2015) 'Synchrony and Exertion During Dance Independently Raise Pain Threshold and Encourage Social Bonding', *Biology Letters*, 11(10), 20150767.

24. Martin, L. J. et al. (2015) 'Reducing Social Stress Elicits Emotional Contagion of Pain in Mouse and Human Strangers', *Current Biology*, 25(3), 326–32.

25. Cohen, E. E. et al. (2010) 'Rowers' High: Behavioural Synchrony Is Correlated with Elevated Pain Thresholds', *Biology Letters*, 6(1), 106–8.

26. Dunbar, 'Breaking Bread'.

27. Kerr-Gaffney, J. et al. (2019) 'Cognitive and Affective Empathy in Eating Disorders: A Systematic Review and Meta-Analysis', *Frontiers in Psychology*, 10, 102.

28. Ibid.

29. Cruwys, T. et al. (2013) 'Social Group Memberships Protect Against Future Depression, Alleviate Depression Symptoms and Prevent Depression Relapse', *Social Science & Medicine*, 98, 179–86.

30. Hamburg, M. E. et al. (2014) 'Food for Love: The Role of Food Offering in Empathic Emotion Regulation', *Frontiers in Psychology*, 5, 32.

31. Martin, L. J. et al. (2015) 'Reducing Social Stress Elicits Emotional Contagion of Pain in Mouse and Human Strangers'.

32. Viero, C. et al. (2010) 'REVIEW: Oxytocin: Crossing the Bridge between Basic Science and Pharmacotherapy', *CNS Neuroscience & Therapeutics*, 16(5), e138–56.

33. Lawson, E. A. (2017) 'The Effects of Oxytocin on Eating Behaviour and Metabolism in Humans', *Nature Reviews Endocrinology*, 13(12), 700–9.

34. Mullis, K. et al. (2013) 'Oxytocin Action in the Ventral Tegmental Area Affects Sucrose Intake', *Brain Research*, 1513, 85–91.

35. Connan, F. et al. (2003) 'A Neurodevelopmental Model for Anorexia Nervosa', *Physiology & Behavior*, 79(1), 13–24.

36. Sharpe, H. et al. (2013) 'Is Fat Talking a Causal Risk Factor for Body Dissatisfaction? A Systematic Review and Meta-Analysis', *International Journal of Eating Disorders*, 46(7), 643–52.

37. Cowdrey, F. A. et al. (2011) 'Increased Neural Processing of Rewarding and Aversive Food Stimuli in Recovered Anorexia Nervosa', *Biological Psychiatry*, 70(8), 736–43.

38. Grzelak, T. et al. (2017) 'Neurobiochemical and Psychological Factors Influencing the Eating Behaviors and Attitudes in Anorexia Nervosa', *Journal of Physiology and Biochemistry*, 73, 297–305.

39. Maguire, S. et al. (2013) 'Oxytocin and Anorexia Nervosa: A Review of the Emerging Literature', *European Eating Disorders Review*, 21(6), 475–8.

40. Lawson, E. A. et al. (2012) 'Oxytocin Secretion Is Associated with Severity of Disordered Eating Psychopathology and Insular Cortex Hypoactivation in Anorexia Nervosa', *Journal of Clinical Endocrinology & Metabolism*, 97(10), e1898–908.

41. Connan, 'A Neurodevelopmental Model for Anorexia Nervosa'.

42. (a) Jones, C. et al. (2017) 'Oxytocin and Social Functioning', *Dialogues in Clinical Neuroscience*, 19(2), 193–201; (b) Hasselbalch, K. C. et al. (2020) 'Potential Shortcomings in Current Studies on the Effect of Intranasal Oxytocin in Anorexia Nervosa and Healthy Controls – A Systematic Review and Meta-Analysis', *Psychopharmacology*, 237, 2891–903.

43. Sainsbury's (2021) 'New Research Reveals Family Dinnertime Is on the Decline with Only 28% of Households Sharing the Same Meal', 12 January. Available at: https://www.about.sainsburys.co.uk/news/latest-news/2021/12-01-21-new-research-reveals-family-dinnertime (Accessed: 28 July 2024).

44. Friend, S. et al. (2015) 'Comparing Childhood Meal Frequency to Current Meal Frequency, Routines, and Expectations among Parents', *Journal of Family Psychology*, 29(1), 136–40.

45. Horning, M. L. (2016) 'Reasons Parents Buy Prepackaged, Processed Meals: It Is More Complicated than "I Don't Have Time"', *Journal of Nutrition Education and Behavior*, 49(1), 60–6.

Chapter 9: Contemplating Contemplative Nutrition

1. Harris, M. (1985) *The Sacred Cow and the Abominable Pig: Riddles of Food and Culture*, New York: Simon & Schuster, p. 165.

2. (a) MacArthur, R. H. and Pianka, E. R. (1966) 'On Optimal Use of a Patchy Environment', *The American Naturalist*, 100(916), 603–9; (b) Harris, *The Sacred Cow and the Abominable Pig*, p. 165.

3. Pyke, G. H. (1984) 'Optimal Foraging Theory: A Critical Review', *Annual Review of Ecology and Systematics*, 15, 523–75.

4. Pontzer, H. (2021) *Burn: The Misunderstood Science of Metabolism*, London: Allen Lane, p. 205.

5. (a) Tishkoff, S. et al. (2007) 'Convergent Adaptation of Human Lactase Persistence in Africa and Europe', *Nature Genetics*, 39, 31–40; (b) Pontzer, *Burn*, p. 204.

References

6. (a) Mathieson, S. and Mathieson, I. (2018) 'FADS1 and the Timing of Human Adaptation to Agriculture', *Molecular Biology and Evolution*, 35(12), 2957–70; (b) Pontzer, *Burn*, p. 205; (c) Rutherford, A. (2020) *How to Argue with a Racist*, London: Weidenfeld & Nicolson, p. 56.

7. (a) Malyarchuk, B. A. et al. (2021) 'Adaptive Changes in Fatty Acid Desaturation Genes in Indigenous Populations of Northeast Siberia', *Russian Journal of Genetics*, 57, 1461–6; (b) Pontzer, *Burn*, p. 205.

8. Pontzer, *Burn*, pp. 198–200.

9. Ibid., p. 206.

10. Online Etymology Dictionary (n.d.) *Culture (n.)*. Available at: https://www.etymonline.com/word/culture#etymonline_v_452 (Accessed: 28 July 2024).

11. (a) Kroeber, A. L. and Kluckhohn, C. (1952) *Culture: A Critical Review of Concepts and Definitions*, Cambridge, MA: Peabody Museum of American Archaeology and Ethnology; (b) Fuentes, A. (2020) *Why We Believe: Evolution and the Human Way of Being*, London: Yale University Press, p. 80.

12. Montanari, M. (2004) *Food Is Culture* (English translation 2006), New York: Columbia University Press, p. 7.

13. Ibid., pp. 51–4

14. Clay, K. et al. (2018) *The Rise and Fall of Pellagra in the American South*. NBER Working Paper No. w23730. Available at: https://www.nber.org/system/files/working_papers/w23730/w23730.pdf (Accessed: 28 July 2024).

15. Billing, J. and Sherman, P. W. (1998) 'Antimicrobial Functions of Spices: Why Some Like It Hot', *Quarterly Review of Biology*, 73(1), 3–49.

16. (a) Ibid.; (b) Sherman, P. W. and Billing, J. (1999) 'Darwinian Gastronomy: Why We Use Spices: Spices Taste Good Because They Are Good for Us', *BioScience*, 49(6), pp. 453–63.

17. (a) Westerterp-Plantenga, M. et al. (2006) 'Metabolic Effects of Spices, Teas, and Caffeine', *Physiology & Behavior*, 89(1), 85–91; (b) Whiting, S. et al. (2012) 'Capsaicinoids and Capsinoids. A Potential Role for Weight Management? A Systematic Review of the Evidence', *Appetite*, 59(2), 341–8.

18. Butler, S. (2018) 'Off the Spice Rack: The Story of Salt', *History*, 22 August. Available at: https://www.history.com/news/off-the-spice-rack-the-story-of-salt (Accessed: 28 July 2024).

19. Pontzer, *Burn*, pp. 188–232.

20. Zaraska, M. (2016) *Meathooked: The History and Science of Our 2.5-Million-Year Obsession with Meat*, New York: Basic Books, p. 39.

Chapter 10: Contemplating Nutrition Science

1. Hwalla, N. and Koleilat, M. (2004) 'Dietetic Practice: The Past, Present and Future', *Eastern Mediterranean Health Journal*, 10(6), 716–30.
2. Osbource, T. B. (1909) *The Vegetable Proteins*, London: Longmans, pp. 1–6. Available at: https://archive.org/details/vegetableprotein00osbouoft (Accessed: 4 August 2024).
3. Hartley, H. (1951) 'Origin of the Word 'Protein'', *Nature*, 168, 244.
4. Funk, C. (1912) 'The Etiology of the Deficiency Diseases', *Journal of State Medicine (1912–1937)*, 20(6), 341–68.
5. Mozaffarian, D. et al. (2018) 'History of Modern Nutrition Science – Implications for Current Research, Dietary Guidelines, and Food Policy', *BMJ*, 361, k2392.
6. Pollan, M. (2006) *The Omnivore's Dilemma: The Search for a Perfect Meal in a Fast-Food World*, London: Penguin, p. 303.
7. Shrimpton, R. et al. (2002) 'Can Supplements Help Meet the Micronutrient Needs of the Developing World?', *Proceedings of the Nutrition Society*, 61(2), 223–9.
8. Funk, C. (2020) 'About Half of U.S. Adults Are Wary of Health Effects of Genetically Modified Foods, but Many Also See Advantages', *Pew Research Center*, 18 March. Available at: https://www.pewresearch.org/fact-tank/2020/03/18/about-half-of-u-s-adults-are-wary-of-health-effects-of-genetically-modified-foods-but-many-also-see-advantages/ (Accessed: 4 August 2024).
9. McIntyre, L. (2021) *How to Talk to a Science Denier: Conversations with Flat Earthers, Climate Deniers, and Others Who Defy Reason*, Cambridge, MA: MIT Press, p. 125.
10. Wunderlich, S. and Gatto, K. A. (2015) 'Consumer Perception of Genetically Modified Organisms and Sources of Information', *Advances in Nutrition*, 6(6), 842–51.
11. National Geographic (n.d.) *Food Staple*. Available at: https://education.nationalgeographic.org/resource/food-staple/ (Accessed: 4 August 2024).
12. Tang, G. et al. (2009) 'Golden Rice Is an Effective Source of Vitamin A', *American Journal of Clinical Nutrition*, 89(6), 1776–83.

References

13. World Health Organization (n.d.) *Vitamin A Deficiency*. Available at: https://www.who.int/data/nutrition/nlis/info/vitamin-a-deficiency (Accessed: 4 August 2024).

14. Balter, M. (2013) 'Farming Was So Nice, It Was Invented at Least Twice', *Science*, 4 July. Available at: https://www.science.org/content/article/farming-was-so-nice-it-was-invented-least-twice (Accessed: 4 August 2024).

15. (a) Mahaffey, H. et al. (2016) 'Evaluating the Economic and Environmental Impacts of a Global GMO Ban', *Journal of Environmental Protection*, 7(11), 1522–456; (b) Zhang, C. et al. (2016) 'Health Effect of Agricultural Pesticide Use in China: Implications for the Development of GM Crops', *Scientific Reports*, 6, 34918; (c) Brookes, G. and Barfoot, P. (2020) 'Environmental Impacts of Genetically Modified (GM) Crop Use 1996–2018: Impacts on Pesticide Use and Carbon Emissions', *GM Crops & Food*, 11(4), 215–41.

16. Examples include: (a) Klümper, W. and Qaim, M. (2014) 'A Meta-Analysis of the Impacts of Genetically Modified Crops', *PLoS One*, 9(11), e111629; (b) Pellegrino, E. et al. (2018) 'Impact of Genetically Engineered Maize on Agronomic, Environmental and Toxicological Traits: A Meta-Analysis of 21 Years of Field Data', *Scientific Reports*, 8, 3113.

17. Shermer, M. (2015) 'Are Paleo Diets More Natural Than GMOs?', *Scientific American*, 1 April. Available at: https://www.scientificamerican.com/article/are-paleo-diets-more-natural-than-gmos/ (Accessed: 4 August 2024).

18. Bloomberg (2021) 'Plant-Based Foods Market to Hit $162 Billion in Next Decade, Projects Bloomberg Intelligence', 11 August. Available at: https://www.bloomberg.com/company/press/plant-based-foods-market-to-hit-162-billion-in-next-decade-projects-bloomberg-intelligence/ (Accessed: 7 August 2024).

19. Shapiro, P. (2018) *Clean Meat: How Growing Meat Without Animals Will Revolutionise Dinner and the World*, New York: Gallery, p. 68.

20. Koop, F. (2021) '5,000 Burgers a Day: World's First Lab-Grown Meat Factory Opens Up in Israel', *ZME Science*, 29 June. Available at: https://www.zmescience.com/science/5000-burgers-a-day-worlds-first-lab-grown-meat-factory-opens-up-in-israel/ (Accessed: 4 August 2024).

21. Believer Meats (n.d.) *Believer*. Available at: https://www.believermeats.com/ (Accessed: 4 August 2024).

22. Koop, '5,000 Burgers a Day'.

23. Watson, E. (2021) 'When Will Cell-Cultured Meat Reach Price Parity with Conventional Meat?', *FoodNavigator USA*, 23 November. Available at: https://www.foodnavigator-usa.com/Article/2021/03/15/When-will-cell-cultured-meat-reach-price-parity-with-conventional-meat (Accessed: 4 August 2024).

24. Tiseo, K. et al. (2020) 'Global Trends in Antimicrobial Use in Food Animals from 2017 to 2030', *Antibiotics*, MDPI. Available at: https://www.ncbi.nlm.nih.gov/pmc/articles/PMC7766021/pdf/antibiotics-09-00918.pdf (Accessed: 16 September 2024).

25. Milburn, J. (2019) 'The Expert Series (2): How Should Vegans Respond to In Vitro Meat?', *The Vegan Society*, 1 February. Available at: https://www.vegansociety.com/about-us/research/research-news/expert-series-2-how-should-vegans-respond-vitro-meat (Accessed: 4 August 2024).

26. Shapiro, *Clean Meat*, p. 24.

27. Little, A. (2018) 'Tyson Isn't Chicken', *Bloomberg*, 25 August. Available at: https://www.bloomberg.com/news/features/2018-08-15/tyson-s-quest-to-be-your-one-stop-protein-shop (Accessed: 4 August 2024).

28. Churchill, W. (1931) *Fifty Years Hence, 1931*. America's National Churchill Museum. Available at: https://www.nationalchurchillmuseum.org/fifty-years-hence.html (Accessed: 4 August 2024).

29. (a) Marsh, N. (2023) 'Why Singapore Is the Only Place in the World Selling Lab-Grown Meat', *BBC News*, 8 June. Available at: https://www.bbc.co.uk/news/business-65784505 (Accessed: 7 August 2024); (b) Skiver, R. (2023) 'Lab-Grown Meat Is Legal in These Countries and Banned in One – a Running List', *Green Matters*', 17 July. Available at: https://www.greenmatters.com/food/where-is-lab-grown-meat-legal (Accessed: 7 August 2024); (c) TOI Staff and Wrobel, S. (2024) 'In World First, Israel Approves Cultured Beef for Sale to the Public', *The Times of Israel*, 17 January. Available at: https://www.timesofisrael.com/in-world-first-israel-approves-cultured-beef-for-sale-to-the-public/ (Accessed: 7 August 2024); (d) *The Economist* (2024) 'How Lab-Grown Meat Became Part of America's Culture Wars', 3 June. Available at: https://www.economist.com/the-economist-explains/2024/06/03/how-lab-grown-meat-became-part-of-americas-culture-wars (Accessed: 7 August 2024).

30. Horton, H. (2024) 'UK First European Country to Approve Lab-Grown Meat, Starting with Pet Food', *The Guardian*, 17 July. Available at: https://www.theguardian.com/environment/article/2024/jul/17/uk-first-

european-country-to-approve-cultivated-meat-starting-with-pet-food (Accessed: 7 August 2024).

31. *The Economist*, 'How Lab-Grown Meat Became Part of America's Culture Wars'.

32. Solar Foods (n.d.) *Food Out of Thin Air*. Available at: https://solarfoods.fi/ (Accessed: 4 August 2024).

33. Huel internal data.

34. PR Times (2021) '完全栄養食品の市場規模、2027年に63億米ドル到達予測' ('Complete Nutrition Market Size Projected to Reach US$6.3 Billion by 2027'), 13 December. Available at: https://prtimes.jp/main/html/ rd/p/000002042.000071640.html (Accessed: 4 August 2024).

35. Pollan, M. (2008) *In Defence of Food: The Myth of Nutrition and the Pleasures of Eating*, London: Penguin, p. 28.

36. Scrinis, G. (2002) 'Sorry, Marge', *Meanjin*, 61(4), 108–16.

37. Scrinis, G. (2013) *Nutritionism: The Science and Politics of Dietary Advice*, New York: Columbia University Press, p. 258.

38. Pollan, *In Defence of Food*, p. 28.

39. Scrinis, *Nutritionism*, p. 258.

40. Ibid., pp. 144–9.

41. Ibid., pp. 238–40.

42. Ibid., pp. 237–8.

43. Britannica (2024) *Food Processing*. Available at: https://www.britannica. com/technology/food-processing (Accessed: 4 August 2024).

44. Revedin, A. et al. (2010) 'Thirty Thousand-Year-Old Evidence of Plant Food Processing', *Proceedings of the National Academy of Sciences*, 107(44), 18815–9.

45. Cablevey (n.d.) 'The Evolution of the Processed Foods Industry', *Cablevey Blog*. Available at: https://cablevey.com/the-evolution-of-the-processed-foods-industry/ (Accessed: 4 August 2024).

46. Hall, R. H. (2000) *The Unofficial Guide to Smart Nutrition*, New York: IDG Books.

47. Monteiro, C. A. et al. (2016) 'NOVA. The Star Shines Bright', *World Nutrition*, 7(1–3), 28–38.

48. Monteiro, C. A. (2009) 'Nutrition and Health. The Issue Is Not Food, nor Nutrients, So Much as Processing', *Public Health Nutrition*, 12(5), 729–31.

49. Van Tulleken, C. (2023) *Ultra-Processed People. Why Do We All Eat Stuff That Isn't Food … and Why Can't We Stop?*, London: Cornerstone Press.

50. Monteiro, 'NOVA'.

51. Numerous meta-analyses; for example: (a) Pagliai, G. et al. (2021) 'Consumption of Ultra-Processed Foods and Health Status: A Systematic Review and Meta-Analysis', *British Journal of Nutrition*, 125(3), 308–18; (b) Askari, M. et al. (2020) 'Ultra-Processed Food and the Risk of Overweight and Obesity: A Systematic Review and Meta-Analysis of Observational Studies', *International Journal of Obesity*, 44(10), 2080–91; (c) Lane, M. M. et al. (2021) 'Ultraprocessed Food and Chronic Noncommunicable Diseases: A Systematic Review and Meta-Analysis of 43 Observational Studies', *Obesity Reviews*, 22(3), e13146; (d) Moradi, S. et al. (2021) 'Ultra-Processed Food Consumption and Adult Diabetes Risk: A Systematic Review and Dose-Response Meta-Analysis', *Nutrients*, 13(12), 4410; (e) Suksatan, W. et al. (2021) 'Ultra-Processed Food Consumption and Adult Mortality Risk: A Systematic Review and Dose-Response Meta-Analysis of 207,291 Participants', *Nutrients*, 14(1), 174; (f) Lane, M. M. et al. (2024) 'Ultra-Processed Food Exposure and Adverse Health Outcomes: Umbrella Review of Epidemiological Meta-Analyses', *BMJ*, 384.

52. Van Tulleken, *Ultra-Processed People*, pp. 52–67.

53. Ibid., pp. 59–60.

54. Visioli, F. et al. (2022) 'The Ultra-Processed Foods Hypothesis: A Product Processed Well Beyond the Basic Ingredients in the Package', *Nutrition Research Reviews*, 1–11.

55. Van Tulleken, *Ultra-Processed People*, pp. 157–9.

56. Ibid., p. 160.

57. (a) Baraldi, L. G. et al. (2018) 'Consumption of Ultra-Processed Foods and Associated Sociodemographic Factors in the USA Between 2007 and 2012: Evidence from a Nationally Representative Cross-Sectional Study', *BMJ Open*, 8, e020574; (b) Juul, F. et al. (2021) 'Current Intake of Ultra-Processed Foods in the U.S. Adult Population According to Education-Level and Income', *Current Developments in Nutrition*, 5(S2), 418; (c) Leung, C. W. et al. (2022) 'Food Insecurity and Ultra-Processed Food Consumption: The Modifying Role of Participation in the Supplemental Nutrition Assistance Program (SNAP)', *American Journal of Clinical Nutrition*, 116(1), 197–205.

58. Cordova, R. et al. (2023) 'Consumption of Ultra-Processed Foods and Risk of Multimorbidity of Cancer and Cardiometabolic Diseases: A Multinational Cohort Study', *The Lancet*, 35, 100771.

59. Hess, J. M. (2023) 'Dietary Guidelines Meet NOVA: Developing a Menu for a Healthy Dietary Pattern Using Ultra-Processed Foods', *The Journal of Nutrition*, 153(8), 2472–81.

60. Gibney, M. J. (2022) 'Ultra-Processed Foods in Public Health Nutrition: The Unanswered Questions', *British Journal of Nutrition*, 129(12), 2191–4.

61. (a) Scientific Advisory Committee on Nutrition (2023) *SACN Statement on Processed Foods and Health*. Available at: https://assets.publishing. service.gov.uk/media/64ac1fe7b504f7000ccdb89a/SACN-position-statement-Processed-Foods-and-Health.pdf (Accessed: 7 August 2024); (b) British Nutrition Foundation (2024) 'British Nutrition Foundation Position Statement on the Concept of Ultra-Processed Foods (UPF)', May. Available at: https://www.nutrition.org.uk/news/position-statement-on-the-concept-of-ultra-processed-foods-upf/ (Accessed: 7 August 2024).

62. Wittgenstein, L. (1921) *Tractatus Logico-Philosophicus* (2001), Abingdon: Routledge.

63. Zaraska, M. (2016) *Meathooked: The History and Science of Our 2.5-Million-Year Obsession with Meat*, New York: Basic Books, p. 168.

64. Birkby, J. (2016) *Vertical Farming*. National Center for Appropriate Technology. Available at: https://attra.ncat.org/publication/vertical-farming/ (Accessed: 4 August 2024).

65. Lynas, Mark (2013) 'Time to Call Out the Anti-GMO Conspiracy'. Cited in McIntyre, *How to Talk to a Science Denier*, p. 132.

Chapter 11: Contemplating What to Eat

1. Watts A. (1951) *The Wisdom of Insecurity: A Message for an Age of Anxiety* (2011 edn), London: Rider, p. 59.

2. Pinker, S. (2021) *Rationality: What Is It, Why It Seems Scarce, Why It Matters*, London: Penguin, p. 5.

3. Fatehi, K. et al. (2020) 'The Expanded View of Individualism and Collectivism: One, Two, or Four Dimensions?', *International Journal of Cross Cultural Management*, 20(1), 7–24.

4. The Economist (2021) 'Technology Can Help Deliver Cleaner, Greener Delicious Food', 28 September. Available at: https://www.economist.com/technology-quarterly/2021/09/28/technology-can-help-deliver-cleaner-greener-delicious-food (Accessed: 4 August 2024).

5. Bartel, S. J. et al. (2020) 'Classification of Orthorexia Nervosa: Further Evidence for Placement within the Eating Disorders Spectrum', *Eating Behaviors*, 38, 101406.

6. The Angry Chef (2016) 'Clean Eating Is Dead', *Angry Chef Blog*, 22 October. Available at: https://angry-chef.com/blog/clean-eating-is-dead (Accessed: 7 August 2024).

7. Gorski, D. (2015) 'The Gerson Protocol, Cancer, and the Death of Jess Ainscough, a.k.a. "The Wellness Warrior"', *Science-Based Medicine*, 2 March. Available at: https://sciencebasedmedicine.org/the-gerson-protocol-and-the-death-of-jess-ainscough/ (Accessed: 7 August 2024).

8. Cancer Research UK (2022) *Gerson Therapy*. Available at: https://www.cancerresearchuk.org/about-cancer/cancer-in-general/treatment/complementary-alternative-therapies/individual-therapies/gerson (Accessed: 4 August 2024).

9. Stearn, E. (2023) 'Dangers of Following a Fruit-Only Diet Revealed After Vegan Influencer "Dies from Starvation and Exhaustion" After Moving to Sri Lanka and Switching to Restrictive Diet', *Mail Online*, 1 August. Available at: https://www.dailymail.co.uk/health/article-12360065/Dangers-fruit-diet-following-death-vegan-influencer.html (Accessed: 7 August 2024).

10. Bratman, S. and Knight, D. (2000) *Health Food Junkies: Overcoming the Obsession with Healthful Eating*, New York: Broadway Books, pp. 235–6.

11. (a) Kinzl, J. F. et al. (2006) 'Orthorexia Nervosa in Dieticians', *Psychotherapy and Psychosomatics*, 75, 395–6; (b) Fidan, T. et al. (2010) 'Prevalence of Orthorexia Among Medical Students in Erzurum, Turkey', *Comprehensive Psychiatry*, 51(1), 49–54; (c) Segura-García, C. et al. (2012) 'Orthorexia Nervosa: A Frequent Eating Disordered Behavior in Athletes', *Eating and Weight Disorders*, 17(4), e226–33.

12. Niedzielski, A. and Kaźmierczak-Wojtaś, N. (2021) 'Prevalence of Orthorexia Nervosa and Its Diagnostic Tools – A Literature Review', *International Journal of Environmental Research and Public Health*, 18(10), 5488.

13. Bartel, 'Classification of Orthorexia Nervosa'.

14. Strahler, J. et al. (2018) 'Orthorexia Nervosa: A Behavioral Complex or a Psychological Condition?', *Journal of Behavioral Addictions*, 7(4), 1143–56.

15. Toti, E. et al. (2019) 'Metabolic Food Waste and Ecological Impact of Obesity in FAO World's Region', *Frontiers in Nutrition*, 6, 126.

16. Serafini, M. and Toti, E. (2016) 'Unsustainability of Obesity: Metabolic Food Waste', *Frontiers in Nutrition*, 3, 40.

17. Toti et al., 'Metabolic Food Waste and Ecological Impact of Obesity in FAO World's Region'.

18. Burton, R. and Sheron, N. (2018) 'No Level of Alcohol Consumption Improves Health', *The Lancet*, 392(10152), 987–8.
19. Nutt, D. (2020) *Drink? The New Science of Alcohol + Your Health*, London: Yellow Kite, p. 58.
20. Burr, M. L. (1995) 'Explaining the French Paradox', *Journal of the Royal Society of Health*, 115(4), 217–19.
21. Fairless, E. (2021) '6 Out of 10 UK Fish Are Being Overfished or Are in a "Critical" State', *Oceana*, 20 January. Available at: https://europe.oceana.org/en/press-releases/6-out-10-uk-fish-are-being-overfished-or-are-critical-state (Accessed: 4 August 2024).
22. Kay, E. (2021) *United Kingdom – Fish and Seafood Market Update*. Foreign Agricultural Service, US Department of Agriculture. Available at: https://apps.fas.usda.gov/newgainapi/api/Report/DownloadReportByFile Name?fileName=United%20Kingdom-%20Fish%20and%20Seafood%20 Market%20Update%202021_London_United%20Kingdom_03-03-2021 (Accessed: 4 August 2024), p. 8.
23. NHS (2022) *Fish and Shellfish*. Available at: https://www.nhs.uk/live-well/eat-well/fish-and-shellfish-nutrition/ (Accessed: 4 August 2024).
24. Marine Stewardship Council (2016) 'UK Government's Healthy Eating Guidelines Recommend MSC Certified Fish', 31 March. Available at: https://www.msc.org/media-centre/press-releases/press-release/uk-government-s-healthy-eating-guidelines-recommend-msc-certified-fish (Accessed: 4 August 2024).
25. Poore, J. and Nemecek, T. (2018) 'Reducing Food's Environmental Impacts Through Producers and Consumers', *Science*, 360(6392), 987–92.
26. Ibid.
27. Ritchie, H. (2019) 'Food Production Is Responsible for One-Quarter of the World's Greenhouse Gas Emissions', *Our World in Data*. Available at: https://ourworldindata.org/food-ghg-emissions (Accessed: 4 August 2024).
28. Poore and Nemecek, 'Reducing Food's Environmental Impacts through Producers and Consumers'.
29. Mottet, A. et al. (2017) 'Livestock: On Our Plates or Eating at Our Table? A New Analysis of the Feed/Food Debate', *Global Food Security*, 14, 1–8.
30. Garnett, T. et al. (2017) *Grazed and Confused?* Food Climate Research Network. Available at: https://www.oxfordmartin.ox.ac.uk/downloads/reports/fcrn_gnc_report.pdf (Accessed: 4 August 2024), p. 16.
31. Ritchie, H. (2020) 'The Carbon Footprint of Foods: Are Differences Explained by the Impacts of Methane?' *Our World in Data*. Available at:

https://ourworldindata.org/carbon-footprint-food-methane (Accessed: 4 August 2024).

32. Reducetarian Foundation (n.d.) *What We Do*. Available at: https://www.reducetarian.org/what (Accessed: 4 August 2024).

33. Pollan, M. (2006) *The Omnivore's Dilemma: The Search for a Perfect Meal in a Fast-Food World*, London: Penguin.

34. Suddath, C. (2008) 'A Brief History of Veganism', *Time*, 30 October. Available at: https://time.com/3958070/history-of-veganism/ (Accessed: 4 August 2024).

35. Montanari, M. (2004) *Food Is Culture* (English translation 2006), New York: Columbia University Press, p. 7.

36. The Vegan Society (n.d.) *Definition of Veganism*. Available at: https://www.vegansociety.com/go-vegan/definition-veganism (Accessed: 4 August 2024).

37. Nguyen, A. and Platow, M. J. (2021) '"I'll Eat Meat Because That's What *We* Do": The Role of National Norms and National Social Identification on Meat Eating', *Appetite*, 164, 105287.

38. Singer (2007), cited by Singer, P. (2020) *Why Vegan?*, London: Penguin, p. 76.

39. Food and Agriculture Organization of the United Nations (2011) *Global Food Losses and Food Waste: Extent, Causes and Prevention*. Available at: http://www.fao.org/3/i2697e/i2697e.pdf (Accessed: 4 August 2024).

Acknowledgements

Several people have given me amazing support and assistance in the research and writing of this book. One friend, however, stands out: Marco Travaglio's input, challenges and critique have undoubtedly made it a significantly better piece of work. His assistance in collating references, helping with phrasing, cross-checking and reviewing has been invaluable. My gratitude is beyond words.

From Huel's Sustainable Nutrition team past and present: thanks to Rebecca Williams, Jessica Stansfield, Jessica Sansom, Matthew Balkin, Madeline Peck, Amy Wood, Emma Detain and Daniel Clarke, for checking various parts, providing insights and/or for helping me with various queries. Thanks to Huel CEO James McMaster for always being a valuable support and sounding-board, and, of course, to Huel founder, Julian Hearn, who, by bringing me in on his brilliant idea, helped to lead me down a contemplative nutrition path.

Huge gratitude to Lazar Vukovic for reading sections and for providing continual reassurance and assistance throughout the writing journey. To Ivan Stefanovic for helping with social media and promotions. Thanks to Nick Dwyer for those many deep debates that have helped to keep me grounded and my biases in check, and to Dave O'Neill for always being a regular ear. Thanks to Drew Price for his collaborations on the UPF debate and for sharing references. To Barbara Bray MBE for inspiring me to actively push back against misinformation. Also to Marta Zaraska for allowing me to pick her brains. To Nerissa Culi for allowing Marco to bounce various ideas along the way. Special thanks, too, to Jason Barnham for more than 20 years of working together in various capacities, the latest being in my activities as a nutrition communicator.

Thanks to Samantha O'Hara of Proofreading Works and Nick Jones of Full Media Ltd for their help with formatting the references. Nick has supported me with my articles for several years and has helped me be a better writer. For the short time we've worked together my agent, Kizzy Thomson, has been an amazing support. Huge appreciation, also, to Simon Alexander Ong for helping me improve the proposal and for introducing me to Kizzy. I couldn't have wished for a better editor: Julia Pollacco has helped to make this book so much better. Thanks also to everyone at HarperCollins who's played a role in editing, formatting, publishing and marketing this book.

Love and thanks to my much-missed mother, Marilynne Collier, whose inspiration and enthusiasm to live a good life helped steer me towards nutrition science. To my dad, Brian Collier, and stepmother, Maureen Collier, for always backing me in everything I do. I'm so lucky to have a father who's never let me down. To Keiran and Chloe for just being themselves. And, of course, endless love to my amazing wife, Mel, who, as well as being a magnificent and innovative cook, continues to support me with everything I do, even when I'm locked away in my study.

Index

Index

Index